Creative Leaps Shape the World

William Graham-Smith

CREATIVE LEAPS SHAPE THE WORLD

The History of the Future

International Books

Graham-Smith, William
Creative Leaps Shape the World; The History of the Future
William Graham-Smith. - Utrecht, International Books. - With index.
ISBN 90 5727 007 2, 224 pages, illustrated
Key words: world history, geology, philosophy, environment

Cover design: Marjo Starink
Printing: Haasbeek

International Books, Alexander Numankade 17, 3572 KP Utrecht, The Netherlands
tel +31 30 2731840, fax +31 30 2733614, e-mail i-books@antenna.nl

CONTENTS

Chapter 1

THE BEGINNING

It is believed that the Universe came into existence about 15,000 million years ago. It is known with greater certainty that it created our Solar System from one small region within itself about 4600 million years ago. There was therefore a considerable interval of time between these two events. During this period the Universe underwent various changes which, among other things, made possible the birth of the Earth and also largely determined its subsequent nature. Any attempt to understand the planet that in due course came to mould our own human nature, and is now the place on which we live, should therefore logically begin at the beginning, and thus with the origin of the Universe itself.

Origins imply a break with what has gone before, and are therefore often difficult territory, and this applies particularly in the case of the Universe. The view that currently prevails stems largely from discoveries made by astronomers and physicists during the last seventy years. They have found that our Sun is merely one of an immense group of stars—there are about 100,000 million of them—that collectively form what is called our galaxy. They also found that this galaxy was merely one of a very large number—again about 100,000 million—of essentially similar galaxies that are separated from one another by vast expanses of almost empty space. In addition there was the very important discovery that the distances between these galaxies are increasing at a uniform rate; in other words, the Universe as a whole is at this present time expanding in all directions in a uniform manner, and so in much the same way as a balloon that is being inflated. It follows that if this expansion had also been taking place in this same manner in the past, then the Universe itself would have been smaller as one passed backwards through time, and would indeed have come into existence out of nothing at a length of time ago that can be calculated. Immediately after its origin it would have been exceedingly small and compact.

This may seem a highly unlikely story, but so would any other designed to account for the origin of something as remarkable as the Universe. The assumption that it has been going on forever, and therefore had no origin, is equally unsatisfactory. The present supposition has the merit of accounting for the expansion that is known to be taking place at the present time. It also apparently fits well with many of the findings of physicists concerning the fundamental nature of the Universe. In addition, about thirty years ago a chance observation led to the discovery that a very uniform low-grade type of energy exists everywhere; this can be regarded as a relic, now almost infinitely cold, of that same energy that was responsible for the original explosive origin of the Universe. This discovery thus provided important further evidence in support of this line of thought.

The present tentative supposition is therefore that our Universe originated suddenly, with explosive energy and so-to-speak from nowhere and out of nothing; this resulted, among other things, in the coming into existence of both time and space. This postulated event is often referred to as the 'Big Bang'. At about 10^{-35} seconds (i.e. about 0.000000,000000,000000,000000, 000000,000001 of a second) after this initial event the equivalent of the whole present content of the Universe was contained within a volume that was about the size of a pea. To us humans this period of time of course seems so incredibly brief that nothing much could possibly happen during it. In fact however the overall conditions then would likewise have been equally extreme. Because such an inconceivable amount was compressed within so small a volume its content would have been unimaginably hot, its temperature being of the order of 10^{28} degrees. The inconceivable degree of activity that would therefore have been taking place in constituents that were inconceivably close to one another could perhaps have resulted in an inconceivable amount of change taking place in the inconceivable short period referred to. In any case it is believed that much fundamental change did in fact take place during this first minute fraction of a second. However the conditions existing at that time would have been so completely different to anything within the experience of humankind that it is apparently impossible for us to know what did take place. And the conditions that gave rise to the Big Bang itself are of course even more fundamentally unknowable.

However this newborn Universe was very rapidly creating further space and expanding into it. Its content therefore became less concentrated, and hence less hot. When it was about one hundredth of a second old some of the photons of radiant energy that comprised the fire ball were becoming converted into matter, in accordance with Einstein's equation $E=mc^2$; here E stands for energy, m for mass and c for the velocity of light. Such relatively familiar particles of matter as protons, neutrons and electrons were then being formed. These would have been existing within an environment consisting of unchanged radiant energy. However little is also known about the changes that took place in the so-called 'quark soup' that existed until the end of the first second. The temperature would by then have fallen to about 10^{10} (i.e. about 10,000 million) degrees, and so had become too low for further photons of radiant energy to be transformed into particles of matter. Thus the overall scene would by this time have been becoming more recognizable.

This Big Bang interpretation therefore implies that the Universe began by what is called a singularity. This means that its nature and the regularities of its behaviour had no antecedents, and therefore cannot be explained as derived from conditions that had previously existed. Indeed before the Big Bang took place time and space had apparently not existed, and precedence therefore had no meaning. It follows that the Big Bang itself and its immediate products, which can here be regarded as including the protons, neutrons and electrons and the gravitational and other influences that they exert, as well as such features as space, time, and the expansion of the Universe, need to be taken for granted. The events that led up to them and 'caused' them remain largely unknown; indeed some of them seem to be in principle unknowable and so may always evade all human understanding. They must therefore just be accepted as 'given'. On the other hand once this initial Universe had cooled somewhat, it developed in systematic ways that are, to a large extent, accessible to human understanding. It is this latter more settled understandable phase that forms the subject of the remainder of this book.

Chapter 2

BEFORE THE EARTH

It therefore appears that after about one second in the life of the Universe the non-understandable endowment phase, which has to be taken for granted, was replaced by a more understandable one which has continued right up to the present time. One can regard this transition as beginning at the stage when protons, neutrons and electrons had recently come into being, and so at a stage when the Universe, as yet only about one second old, was still quite small and intensely hot. This present chapter is concerned with the developments that were eventually, some ten thousand million years later, to make possible the formation of the Sun along with the Earth and other planets of our Solar System.

The situation with regard to these protons, neutrons and electrons is interesting. An important feature is that they exerted influences, or 'forces', that extended outwards into surrounding space, and thus well beyond the confines of these very small particles of matter themselves. There are four such types of forces, and two of these, namely gravitation and electromagnetism, are fairly familiar in our daily lives. In addition, each proton is associated with a single positive electric charge, and each electron with a comparable negative one. Neutrons have almost exactly the same mass as protons, but differ in having no charge. Electrons are very much smaller and less massive than protons or neutrons.

At the very high temperatures involved, these newly-created material particles were moving about in all directions at high speeds. If two protons approached one another, because each of them carried a positive electric charge, they mutually repelled one another. However if one was moving directly towards another particularly rapidly, then they were liable to break through this repulsion barrier and become so close together that another of the four 'given' forces, namely the strong nuclear one, came into play. These nuclear force were mutually attractive, and far more powerful. The outcome of dynamic activities of this kind was that two protons and two neutrons became closely associated with one another in such a way that by their mutual interplay they collectively formed a new single dynamic and closely integrated system that was larger and more complex than the single protons and neutrons that had given birth to it. At first the intensity of the surrounding energy was so great that these new systems were destroyed almost as soon as they were created. Their existence was therefore merely transient. However the Universe was continuing to expand, and its content—therefore—to cool, and after about 100 seconds the surrounding environmental temperature had fallen to 1000 million degrees. Under these cooler conditions these new systems were no longer merely transient. They were stable, and indeed highly sustainable, for systems created at that time are still functioning now some fifteen thousand million years later. Thus developments were demonstrating that already at this very early stage in the history of the Universe two particle systems could mutually interact with one another in a way that led to the creation of a new third system that was also sustainable, and was larger, more complex and with a different nature to that of either of its parent systems. An inherent creativity that gave rise to increasing diversity and complexity was thus already in existence at a very early stage in the history of the Universe.

The expansion and consequent cooling of the Universe continued. After about 300,000 years, and by then at a temperature of about 3000 degrees, conditions had become sufficiently tranquil to permit an electron that came into the vicinity of a proton to remain in a sustainable longterm relationship with it, based on the mutual attraction between their dissimilar electric charges. In this case the proton and the electron remained at a relatively great distance from one another, but nevertheless by their mutual interactions they created a single integrated system of a type known as an atom. This situation is to some extent comparable to the way in which our Sun and its planets mutually interact despite their distance from one another and collectively form the Solar System. In the case of atoms it is their proton/neutron component that, being the much more massive portion, lies at the centre of the complex and is referred to as its nucleus, and it is electrons that form the peripheral components. In the simplest type of atom, namely that of hydrogen, the nucleus consists of a single proton, and the peripheral partner is a single electron. The positive charge on its proton and the negative one on its electron cancel one another, so that overall the atom is electrically neutral. In a similar way the system comprising two protons (each with one positive charge) and two neutrons (each with no charge) becomes the nucleus of an atom that has two peripheral electrons and is therefore also electrically neutral. This is a helium atom. For a very long time these were virtually the only types of nuclei, and hence also of atoms, that had come into existence.

Most of the hydrogen atoms soon organised themselves into pairs. Their nuclei (in this case each a single proton) remained separate from one another; the two electrons were, again, peripheral. The resulting products, namely hydrogen molecules, also behaved as single, integrated and dynamic systems. Molecules were therefore another new creation which, again, was larger and more complex than the atoms which preceded them and which had created them as a result of their mutual interactions.

At this stage there was a Universe that was some 300,000 years old and had, by its continuing expansion, created quite a large volume of space. This space housed radiant energy that was ultimately derived from the Big Bang. Within this environment there was a tenuous gas consisting mainly of hydrogen molecules along with a smaller number of helium atoms. This Universe would then have seemed a rather desolate place, monotonous and homogeneous, the only obvious change being that it was itself continuing to expand, and that its contents were therefore becoming ever colder and more tenuous. This uninspiring state of affairs could presumably have continued indefinitely had not one of the initial 'given' forces, namely gravitation, come to gradually transform the scene.

Gravitation draws all material entities towards one another. If, then, after the Big Bang there had been some small irregularities in the overall scene, such that certain regions were rather more dense than others, the consequences would be important. Certainly the mass of a hydrogen molecule is exceedingly small, but nevertheless if these molecules happen to be rather more numerous in a particular region of space, then there will be a tendency for outlying molecules to gravitate towards this region, further increasing the concentration of molecules there. At first this process would have been exceedingly slow, but its rate would have gradually increased because the greater the concentration of gas the greater its power to draw increasingly more gas towards itself. This continuing self-reinforcing or positive feedback situation would come to result in the creation of large numbers of huge localised regions of more concentrated gas, separated by other regions that were comparatively devoid of matter. We could also expect that within these more concentrated regions more localised areas with even greater concentrations would develop.

Further results in these last localised regions would have been impressive. The hydrogen near the centres of such areas would have become progressively denser, and therefore also hotter, owing to the weight of the peripheral gas attracted towards these foci by the force of gravitation. The increasing temperature and associated motion first caused the hydrogen molecules to part company, returning to their earlier state of separate atoms. Then, at still higher temperatures, these atoms were stripped of their peripheral electrons, once more giving rise to protons that were no longer surrounded by peripheral electrons. Still later a temperature was reached at which protons plus neutrons became converted into helium nuclei in the same manner as had occurred at a comparable temperature at an earlier stage in the history of the Universe (p.12). This transformation was important not so much on account of the helium nuclei that were produced, but rather because the mass of a helium nucleus is slightly less than that of the protons and neutrons from which it has been created. This difference in mass is transformed into a great quantity of energy in accordance Einstein's famous $E=mc^2$ equation.

It is probable that developments of this kind led to the formation of the large-scale concentrations of matter that became galaxies, and also to the presence within these galaxies of the more localised and denser clouds of gas and dust within which large numbers of stars are at present coming into existence. Astronomers using recently invented telescopes that make use of infrared radiation, can now see into the interior of these clouds of gas and dust, which are transparent to this type of radiation. Another aid to the study of these clouds are satellites that can raise these telescopes high enough above the Earth to avoid the complications that would otherwise be caused by our own planet's atmosphere. It appears that, at a critical stage, these clouds of gas and dust collapse at certain points owing to force exerted by their own weight. This collapse raises the temperature so greatly that nuclear fusion processes that transform hydrogen nuclei into helium nuclei are initiated, along with the huge output of energy associated with this process. Furthermore the radiation that is emitted during this process clears away the surrounding gas and dust. Not only do these circumstances create a star, but also as a result of its birth the surrounding region is cleared of gas and dust, allowing the newborn star to be visible from far away.

Embryo stars are associated with enough gas and dust to enable planets to be formed. Initial slight irregularities cause these clouds of gas and dust to begin to rotate, and as material continues to fall inwards the speed of this rotation is increased, for much the same reason as skaters rotate more rapidly when they draw their arms inwards. It is likely that numerous proto-stars have clouds of gas and dust circulating external to themselves and in the same plane as their equators. It is out of such circulating material that the planets of our

own Solar System have condensed. We do not at present know if this formation of planets has also occurred in relation to other stars, for even the closest
of them is at such a great distance from our own star, the Sun, that it is still
impossible to know for certain whether or not they also have planets revolving
round them.

A star, then, is an aggregate that consists mainly of vast numbers of separate hydrogen nuclei (i.e. protons). It is a very different kind of system to the
nucleus of an atom, or an entire atom, or a molecule, which are each composed
of a small number of mutually interacting units. All three of these are highly
dynamic integrated systems but, being closed ones of a type known as standing
waves, the latter need neither to take in nor to emit energy; they can, given favourable circumstances, just go on functioning as integrated entities more or
less forever. The aggregation that comprises a star, however, is an open system, which can continue to function only if there is a continual flow of energy
passing through it. The input is nuclear energy fuelled by the transmutation of
hydrogen nuclei into helium nuclei; the output is the heat and light radiated
outwards into the space around the star. This results in the depletion of the
star's stock of fuel, namely hydrogen nuclei. A star, unlike atoms and other
closed systems, cannot just go on functioning without change forever. It has to
have a history.

What, then, is the history of a typical star? It starts as a stable organization.
The pressure generated at its centre by the nuclear fusions supports the
weight of the overlying material; an increase or decrease in this weight will in-
duce a corresponding change in the rate of nuclear fusion. This maintains a
balance. Similarly increased fusion at the centre increases the rate at which energy percolates to the exterior of the star and then is radiated into space, so in
this respect also a balance is maintained. The whole system involves an expen-
diture of hydrogen nuclear fuel. It will therefore inevitably cease, or at least
change drastically, when this fuel becomes depleted. Yet a typical star continues to function in a stable state for an exceedingly long period. Our Sun
was formed about 4.6 thousand million years ago, but as yet only about one
half of its available nuclear fuel has been used. The Sun can therefore be expected to continue in its present stable state, and to radiate much the same
quantity of heat and light, for another three or four thousand million years.
This, at least by human standards, is a very long time.

All such systems, including our Solar System, must eventually come to an
end. When the depletion of hydrogen nuclei at its centre reaches a certain
stage the star becomes unstable, for the pressure generated there is no longer

sufficient to support the weight of substance pressing down on it. A partial collapse inwards ensues. The resulting sudden compaction at the centre of the star generates temperatures higher than any it had previously experienced; these initiate a nuclear reaction in which helium (not hydrogen) nuclei generate the synthesis of new and more complex types of atomic nuclei, in particular carbon, which has 6 protons and usually 6 neutrons in its nucleus, and oxygen, which has 8 protons and usually 8 neutrons. In stars which are not much larger than the Sun there is usually little further change. Old age sets in, and the stars gradually cool down, becoming degenerate structures known as 'white dwarfs'.

With stars that are more than eight times the size of the Sun, the terminal part of their life-histories is more dramatic. A series of partial collapses continues beyond oxygen, but at shorter intervals since less energy is generated at each successive stage. Finally with the nuclei of iron (26 protons and 30 neutrons) the amount of energy generated is reduced to zero. The result is catastrophic. With the failure to maintain the internal pressure, the vast mass of the whole external portion of what is an exceptionally large star falls, virtually instantaneously, into its central region. This sudden total collapse generates such a colossal temperature and so much radioactivity that the whole range of more complex atomic nuclei, from iron with its 26 protons onwards to uranium with its 92, are synthesised in various quantities. Also, importantly, this implosion of the whole exterior of the star causes a catastrophic rebounding explosion to generate in the star's interior; this is so extreme that the star itself is largely destroyed and much of the material content of its outer layers is ejected with tremendous violence into surrounding space. The energies involved are such that a star which could previously be seen only through a powerful telescope can become, for a short time, so bright as to be easily seen by the unaided human eye. A star that undergoes such a dramatic explosion is known as a supernova.

Supernova explosions are rather uncommon. In our own galaxy, for instance, the last ones were in 1572 and 1604. There was one in an adjacent galaxy in 1987, which was near enough to provide astronomers with an immense amount of important information. Though rare, supernovas have played a very important part in shaping the overall evolution of the Universe. Almost all the existing types of atomic nuclei have been created within such stars, with the exception of hydrogen and of most of the helium; the relative numbers of these types differ widely, for reasons that are understood; those of carbon, oxygen, nitrogen and silicon were for instance numerous, those of gold much less so. Eventually the supernova explosions that followed scattered much of this material into the cool surrounding outer space, where these nuclei became associated with electrons, forming the equivalent types of atoms.

The grains of material produced in this way were small and came to form the dust component in the clouds of gas and dust already mentioned. The quantity of such dust has presumably been increasing as supernovas of succeeding generations of stars each discharged their own contribution. In the contracting cloud of gas from which our Sun and its planets was formed there was apparently sufficient of this dust to contribute about one percent to the material content of these bodies. The other 99 percent consisted of hydrogen and helium. Nevertheless it is this one percent that has given rise to almost all the types of elements that are present both in the Sun and in its planets. It is this same one percent that has caused planet Earth to be largely solid, instead of merely a ball of gas. And, lastly, it is also this one percent that has made possible the emergence and subsequent evolution of living organisms, including ourselves, on the planet on which we are now living.

This Universe is thus a quite remarkable phenomenon. Its mysterious beginning provided some initial 'given' systems, along with their qualities and procedural rules. It has created, from its raw material, a succession of new kinds of things and scenes, along with appropriate new qualities and new environments. The origin of each development depended on the earlier existence of its predecessor. On this ongoing basis it created first helium nuclei, then hydrogen and helium atoms, then hydrogen molecules, and then the concentrations of hydrogen molecules that led eventually to stars, then within some of these stars the creation of many new types of atomic nuclei, then the scattering of these into surrounding space by supernova explosions, then the conversion of these nuclei into new equivalent atoms, and the incorporation of some of the resulting dust into new generations of stars and whatever planets are associated with them. On planet Earth there has been a further succession of creative events; these have been taking place in a cool isolated planetary environment and have involved numerous rather gentle chemical reactions instead of the violent nuclear ones that are characteristic of most portions of this Universe.

Spiral galaxy

Chapter 3

THE MAKING OF THE EARTH

Our Earth, along with the other planets of the Solar System, was formed about 4600 million years ago in the course of the creation of one particular star, the Sun. One can try to visualise our proto-Sun spinning round, along with a disc of tenuous material extending outwards from its equator for hundreds of millions of miles. This disc, which was rotating in the same plane as the star itself, would have consisted principally of hydrogen and helium, but it would also have included a sample of the ninety or so types of atoms whose nuclei had been synthesised in the interiors of earlier generations of stars. The deep interior of our proto-Sun was exceedingly hot; nuclei of atoms could exist there, but not the atoms themselves. On the other hand the disc was relatively cool; there electrons could not only become associated with atomic nuclei to form atoms, but the dynamic activities that resulted could be maintained indefinitely; in other words, these atoms were sustainable. Also it was cool enough for some of them to interact with one another in ways that led to the formation of molecules, and these also were sustainable. The events leading to this situation were indicated in the previous chapter.

Atoms are also present on the Earth, and are indeed the material from which it itself and everything thereon has been created. They are therefore very important. Also, being present in abundance, they are readily available for study. A few comments on what is known about these atoms, and also about the molecules that are formed from them, may be helpful at this stage. As already noted (p.12), the number of electrons that become associated with an atomic nucleus is equal to the number of protons in that nucleus. The resulting atom is therefore electrically neutral. These electrons remain at a relatively long distance from the nucleus. Their distribution is by no means random, for they are located along circular paths, known as orbitals, that are at various successive distances from the central nucleus. The electrons are restricted to these orbitals by the resonances of wave-systems that are part of these integrated atomic systems. One might say that this cloud or swarm of external electrons has a cushioning or moderating influence, for when two atoms come into contact and interact with one another it is not their nuclei which are shielded, but these peripheral electrons that are involved in the interactive processes. These chemical interactions are therefore electronic, not nuclear, and the energy associated with them is much less intense and violent.

It is important that each orbital can afford accommodation for only a limited number of electrons; in the case of the innermost orbital this number is two. and for the next it is eight; for subsequent orbitals the situation becomes more complicated. Each successive orbital has to be fully occupied before the occupation of the next one further outwards can begin. An atom in which the outermost orbital is fully occupied is chemically very inert. It is, for the atom in question, the most satisfactory or 'preferred' state. Helium, for example, has only one orbital, namely the innermost, and it has its full compliment of two electrons. Helium atoms are therefore chemically very inactive.

Other types of atoms play a more active role. Like helium, the hydrogen atom has only one orbital. It accommodates only one electron; greater stability would require a second electron, but then the system would no longer be electrically neutral. An atom of oxygen has eight protons, and therefore also eight electrons. Two of these fill the inner orbital, and the remaining six are in the next orbital, which would be fully occupied if it had eight. If hydrogen and oxygen come into contact, a major re-organization takes place so rapidly as to be explosive. The single electron in each of the hydrogen atoms obtains, in effect, a partial accommodation in one of the two vacancies in the outer orbital of an oxygen atom. These two electrons are shared by hydrogen and oxygen atoms. The outer orbital of the oxygen atom obtains its full compliment of eight electrons, though only at the cost of two of these being shared with hydrogen atoms. Similarly, this process fills the single hydrogen orbital. Linkages of this type, based on the sharing of electrons, are frequent. They draw the atoms concerned into a new form of integrated organization, namely a molecule, in the case of hydrogen and oxygen into a molecule of water. However the crucial point is that the structural organization, and therefore also the internal dynamic activity, of the molecule is different from that of either of the parent atoms from which it is derived. In short, a total integrated entity or system of a new and more complex kind has been created. Because the overall infrastructures are different, the characteristics and qualities of the corresponding substances are likewise different. In the case of oxygen and hydrogen the resultant material, namely water, clearly differs in almost all respects from either of the parent materials. A new system, and consequently a new kind of material substance, has been created as a result of the chemical reaction.

Such re-organizations provide a principal means whereby new kinds of substances have come into existence. The atoms themselves, with their limited structural potentialities, normally permit the existence of only ninety-two types of elements on the Earth. Chemical reactions between these elements make a far larger number of types of molecules, and therefore of molecule-based compounds, possible. This creates the potential for a considerable amount of material diversity. Nevertheless, for highly complex reasons related to orbitals, resonances and so forth, most types of atoms are only able to participate in the formation of a rather limited number of kinds of compounds.

Importantly, however, atoms of the element carbon are in an exceptional position. Their second orbital has four electrons. They can therefore reach a more satisfactory electron arrangement in numerous different ways. In addition, also importantly, these carbon atoms can readily adjust by sharing electrons with one another. There is scope for the formation of an almost infinite number of large and complex types of molecules that are held together by chains, or other configurations, of interlinked carbon atoms. It is complex carbon-based molecules of this type that have, among other things, provided the infrastructure that has enabled living organisms to come into existence on the Earth.

There is another aspect of atoms that is highly relevant to our story. When atoms (or molecules) of the same type come into close proximity, the adjacent electrons of the outermost orbitals of each tend to be mutually attracted to their counterparts; the result of these weak attractions is that these atoms, while retaining their overall individuality, remain adhered to one another. In contrast, there is also a universal tendency, dating from the Big Bang, for such units to be in a continuous state of motion, and the higher the temperature the more vigorous this kinetic motion will be. The adhesive forces and those of kinetic motion thus act in opposition. Where the adhesion forces are strong, as for instance in iron, and where the temperature is what one might call normal, the amount of kinetic motion is not sufficient to disrupt the adhesive forces, and the latter retain the iron in a solid unyielding state. On the other hand the kinetic energy is sufficient to cause the individual atoms to vibrate while still remaining in the same relative positions. Heating the iron somewhat will, by increasing the temperature, increase this vibration, which will in turn cause the heated metal to expand. With continued heating a stage will be reached at which the adhesion force is no longer sufficient to maintain the atoms in their previous relatively fixed positions; the iron will sag under its own weight, and will be said to be melting, so changing from a solid to a liquid state. The adhering tendencies are still sufficient to keep the surfaces of the atoms in close proximity to one another, so that they now flow past one another instead of, as in solids, being retained in the same fixed relationships. Finally, at a still higher temperature, the kinetic movements of the atoms become so vigorous that the adhering tendencies break down altogether. This gives rise to isolated iron atoms whose kinetic energy keeps them moving through space in all directions, colliding with one another, and so forth, the iron having thus become a typical gas.

In the light of these comments we can turn once more to the equatorial disc that is present during the formation of a star such as the Sun. The whole system will be rotating. Both star and disc will be subject to gravitational attraction. The material of the equatorial disc, along with everything subsequently formed from it, will already be set on a course that sends it revolving round and round the parent star, which will be at the centre of any resultant planets. Most of the material in the disc will be hydrogen or helium that was created soon after the Big Bang. The remainder, about one percent, will consist of atoms that had their nuclei synthesised deep in the interiors of earlier generations of stars and, after being disseminated by supernova explosions, had come to form, as atoms, the dust in interstellar space.

Much of this disc material will be circulating at a considerable distance from the star, and will therefore be relatively cool. It will, however, be absorbing some of the radiant energy emitted by that star, and will to that extent be warmed. Some types of atoms, such as those of iron referred to above, are likely to adhere to one another, and continuing aggregation will eventually lead to the formation of small localised solid masses of such materials. Other elements that had little tendency to adhere, such as oxygen, would have been present in the form of gases. Gravitation would have tended to draw smaller bodies towards whatever larger masses first materialised. The larger ones would presently have come to have the status of proto-planets, and the remaining smaller ones would have been liable to crash into them, so becoming early meteorites. Smaller particles derived from interstellar dust, and also materials existing as gases, would have been drawn in by gravitation, so adding to the substance of the developing planet. In this way, the material of the former disc would have been increasingly transformed into a few comparatively large bodies, the planets, which would by now be separated from one another by almost empty space. These planets, like the discs from which they were formed, would have been circling round the parent star. It is important that these planets were not small precise highly integrated forms of organization, as were for example atomic nuclei and atoms. On the contrary, they consisted in their totality merely of aggregates of atoms drawn haphazardly together.

Probably all planets have much in common, but our attention here is centred on the one that particularly concerns us humans, namely our Earth. At this early stage, the aggregation of materials of which it was composed would have been a busy scene of interactions and transformations, all occurring more or less simultaneously. Many different types of atoms would have been present in close proximity to one another, along with the new incoming material contributed by meteorites. The whole conglomeration would have been kept warm partly by the Sun's radiation and partly by heat arising from disintegrations of certain types of unstable radioactive atoms. The young planet Earth was at this stage a rather warm and hectic place where chemical interactions of numerous different kinds were giving rise to many new kinds of molecules, and where much solidification, melting, vaporization, crystallization and so forth was taking place. It amounted to a vigorous stirring, sorting and developing of the materials that had come to hand.

One result of this activity was that the substance of the Earth was gradually transformed from a fairly chaotic mix into a somewhat structured place. Iron was abundant and heavy; nickel was likewise heavy. Hence, under the influence of gravitation, these two heavy metals gravitated towards the centre of the Earth. This central region became particularly hot, for it is there that the heat generated as a by-product of radioactivity had the least opportunity to escape. Much of this metal therefore was, and still remains, in a molten state. The lighter materials, such as the aluminium-silicon-oxygen compounds that were formed in great quantities, rose upwards to the surface of the mass. Here the temperature was much lower, and the melting points of these compounds were high enough for them to solidify; they formed, and still form, a relatively light, but solid, crust. The temperature in the intermediate region between this core and crust is also intermediate. This material, which is known as the mantle, is on the one hand subjected to pressure from the weight of the crust above it, which tends to keep it compact, dense and solid, and on the other hand it is subjected to heat of radioactive origin which tends to keep it molten. It is consequently neither liquid nor solid, but is in a hot dense viscous state in which slow convection movements are taking place.

Owing to the high temperatures prevailing in the interior of the Earth, great quantities of water molecules in the form of steam were created from the hydrogen and oxygen. The pressure exerted by the weight of the overlying materials forced this gaseous steam to push its way up to the surface. At the lower temperature at the surface the steam became transformed into liquid water. This water flowed downwards and accumulated in major depressions in the surface of the Earth's crust, and formed the first oceans.

Other gases, such as hydrogen, helium, nitrogen and carbon dioxide, similarly made their way to the surface, but in these the self-adhesive tendencies were not strong enough to make them liquefy at the temperatures they encountered. They remained in a gaseous state and collectively formed an atmosphere that enveloped the whole Earth, covering the surfaces of both land and sea alike. Much of the hydrogen and helium, being light, diffused upwards beyond the planet's gravitational control, and escaped into outer space. Hence the Earth's early atmosphere probably came to consist largely of carbon dioxide and nitrogen.

This account may give the impression that these various structural relationships are essentially static, and this was in fact the accepted viewpoint until fairly recently. Doubts on this score first became apparent when, early in the present century, Alfred Wegner assembled evidence indicating that about 300 million years ago the geological structures, the plants and the climate on the two sides of the Atlantic Ocean had been remarkably similar. He suggested that at that time there had been a single super-continent, and that subsequently its Euro-African and its American parts had drifted away from one another to reach their present positions. His concept of continental drift was not widely accepted, mainly because it was difficult to conceive a mechanism that could have induced whole continents to move in this kind of way.

A number of exciting developments occurred in rapid succession from 1950, due largely to the application of more refined techniques. It was found, for instance, that those minute particles in molten lava that happen to be sensitive to magnetism orient themselves in the direction of the Earth's magnetic field; they then remain permanently 'frozen' in this position when, on cooling, the lava solidifies. It follows that a palaeo-magnetic study of rocks formed in different places and at different times can be used to ascertain the direction of the Earth's magnetic field at the times in question. If this was done this using rocks of various early ages on one side of the Atlantic Ocean, a curve showing magnetic polar wandering was obtained. If it was done on the other side, then a similar curve was obtained, but apparently at a different location. If however the calculation of the pole's positions was made on the assumption that the American and Euro-African continents had been in apposition, instead of separated by an ocean, then the two curves approximately coincided. Since the Earth cannot have two sets of magnetic poles this evidence strongly suggested that during the earlier period these two continents had in fact been contiguous. And this of course implied that they had subsequently drifted apart.

Studies of residual magnetism also led to the surprising discovery that, in some cases, the direction of the Earth's magnetism was the reverse of what had been expected. These studies aslo showed that the same reversal occurred in all rocks of the same age. It became clear that periodically (once or twice every million years) the polarity of the Earth's magnetic field has been reversed.

By using sonar and echo location, researchers in the 1950s demonstrated that great mountain ridges straddled the deep ocean floors. In the Atlantic Ocean a ridge some two miles high extends continuously from Iceland in the north to Tristan da Cuhna in the south; the intermediate zone lies below a great depth of ocean water. Along the central axis of this ridge is a deep depression that resembles the Great Rift Valley in Africa; it is about one mile deep and thirty miles across. The mountain ridges on each side of it are composed of lava. It was also found that the direction of the fossil magnetism in this lava is periodically reversed; the rocks on each side of the depression show a series of reversals that run parallel with it.

Results from these very different lines of research dove-tailed neatly. Recent work has indicated that where the Earth's crust has fractured, as for instance in the rift-like valley that traverses the Atlantic Ocean, the pressure on the underlying mantle is reduced. This causes some of the adjacent mantle material to melt, and hence to rise up as molten lava. Here it became magnetised in accordance with the direction and polarity of the Earth's magnetic field at that time, and this position was retained after the lava had cooled and solidified. This newly-formed rock was then moved in a horizontal direction to one side or the other of the rift as further molten material rose up and took its place. Periodic reversals of the Earth's magnetic field cause reversals in the direction of the magnetism in the resultant igneous rock, and because this has subsequently moved horizontally away from the rift, a series of parallel strips of rock that have their residual magnetism oriented in opposite directions has resulted. The dates of recent reversals are known from observations made elsewhere. This makes it possible to calculate the approximate rate of lateral movement. In the Atlantic the rate of continental drift is between 1.5 and 2.0 centimetres per year (this rate has recently also been measured directly); elsewhere there are places where this rate of drift is as much as 7.0 centimetres per year.

This manner of lava production and solidification has led to the formation of rigid masses of rock that, though relatively thin in a vertical direction, are often of great horizontal extent. These are known as tectonic plates. They are being propelled over the surface of the mantle at a very slow but continuous rate. The initial Euro-African-American continental block split into separate Euro-African and American parts, which were then subsequently transported horizontally on a divergent course. Wegner's supposition had been correct; there was—and still is—a mechanism that can move whole continents horizontally. As regards the Atlantic region, the upwelling rift valley system that has been engineering the divergence of the two continental blocks now extends from north to south midway across the ocean of water that has filled the trough left as a result of these two continents drifting apart. This rift is at present a site of earthquakes and volcanic activity, as is seen more particularly where its northern end, as Iceland, surfaces above the water. The associated volcanoes, earthquakes and geo-thermal phenomena can be easily studied there.

It is not surprising that these massive plates, moving over the surface of the mantle along a horizontal plane, should sometimes collide or become entangled with one another. Regions of contact can become an additional scene of earthquake and volcanic activity, and sometimes also of crustal mountain building. The San Andreas fault in North America is a well-known example; here two tectonic plates, moving on different courses that impinge obliquely, was recently responsible for major earthquakes near San Francisco and for the renewed eruption of the volcano at Mount St Helen. Similarly the plate associated with the Indian subcontinent is moving northwards and making slow but violent contact with an Asian plate; the result has been a slow but continuing intense buckling of crustal rocks that has resulted in the gradual building of the huge Himalayan mountain range. On occasion, one plate can slide under another, and the lower one can then become re-incorporated into the substance of the mantle. In the very long-term the material of the mantle is therefore circulating, or being recycled; it not only wells up from the mantle but also, after a temporary transformation into tectonic plate material, returns to the mantle.

Until comparatively recently the distribution of continents and the existence of mountain ranges, earthquakes and volcanic activity were still seen as disconnected and largely inexplicable phenomena; the tectonic plates were unknown and the drifting of continents discounted. Today all these are viewed as aspects of a single inter-related pattern of activity. The energy required to drive this vast system with its immense inertia is ultimately derived from the disintegration of radioactive atoms embedded in the planet's core and mantle.

The Earth sciences have thus become more unified, more explicable, more interesting and also far more complex; they have, in short, taken a major step forward.

The Earth's atmosphere also moves, and, being composed of gases, it is of course vastly more mobile than the mantle. In this case it is mainly energy radiated from the Sun that fuels the system. The regions nearest to the Earth's equator receive the most heat. This is therefore where the evaporation of water from the surface of the sea is most rapid, and where the strongest convection currents are generated in the adjacent air. This warm rising air carries large quantities of water, as vapour, high into the sky and there, in this colder environment, the vapour that it has been carrying condenses into innumerable minute droplets of water that collectively give rise to clouds. The equatorial region near sea-level thus left relatively devoid of air is occupied by cooler air that is drawn towards the equator from more temperate regions some 30 degrees further to the north or south. This air is in turn replaced by air that, after rising in the tropics, has moved north or south at higher altitudes. In this way the radiation from the Sun sets up a system of circulating air in which, in principle, winds blow towards the equator at low altitudes, and away from it at higher ones. In practice the effects of this primary driving force are greatly modified by the rotation of the Earth, by seasonal changes in the relative position of the Sun, by differential rates of heating and cooling on land as compared with the sea, and by the existence of mountain ranges. All this complicates the way in which air circulates; in many regions the weather will often vary considerably from day to day.

The water that has condensed to form clouds will fall as rain. The part that falls on land will run downwards, either on its surface or through pores in the rocks below, and will eventually gather, first into streams and then into rivers, and make its way back to the sea. Water therefore circulates. The streams will detach and carry downwards small particles of rock and debris from the mountains and lower foothills. This material will later be deposited as sediments where the flow of water has abated, perhaps in lakes, or in the plains, or in the sea. Thus great mountain ranges are not only slowly raised as a result of plate tectonics but they are also slowly eroded as a result of atmospheric processes.

In short, the Earth's mantle, its atmosphere and its water are all in various states of motion. Some of these processes interact with one another, and overall they form a stable yet dynamic planetary system that has probably continued to operate essentially unchanged ever since the early chaotic genesis of the Earth had been completed some four thousand million years ago. There have, however, been important differences in the way in which this overall system has expressed itself at different periods, due largely to the unyielding character of solids. The drift of a continental block to a new location is a substantial long-term change, and a new mountain range will not be quickly eroded away. There have therefore been long periods when particular parts of the surface of the Earth have been hotter or colder than usual, or wetter or drier, and so forth. Thus although the Earth's interior may not have changed much, its external surface, which has provided life with its environment, has been continually changing, though not apparently in any one particular direction. This surface region has been a changeful place, very different to its near neighbour the Moon. Principally because the Moon is so much smaller, it no longer has tectonic plates, atmosphere or water, and is therefore a dead unchangeful place.

Chapter 4

EARLY LIFE ON THE EARTH

By the time the Earth had completed the initial chaotic phase of its development, its physical structure had become essentially similar to that of the Earth today. There were already continents and oceans and an atmosphere. The planet was making a yearly journey round the Sun, and was also rotating daily on its axis. Already there would have been summer and winter, day and night, and also winds, clouds and rain, and likewise streams, rivers and lakes. There was however one great difference. No life as yet existed. No plant or animal was experiencing or contributing to this early planetary scene.

What then is this life that did not at that time exist? It can easily be seen in a relatively simple form by viewing, through a microscope, a little water taken from a greenish puddle when, during the spring, light from the Sun is becoming more intense. Large numbers of minute creatures will be seen propelling themselves through the water by lashing movements of two whip-like projections called flagella. A detailed study would show that each is a sensitive, energetic and self-sufficient living system that is capable of responding to a particular situation and acting as a single whole. Each is also able to grow, to reproduce itself, and to do these things without losing its distinctive features. To see these minute organisms—so small, so full of vitality and apparently of purpose—demonstrates dramatically that life—even very small life—is very different to non-life. The cells seen in the puddle will be eukaryote cells (see chapter 5). Prokaryote cells (i.e. bacteria) are smaller and simpler, and were evolved earlier. It is with prokaryote cells that this chapter deals.

The familiar plants and animals that we see around us, and we ourselves, are of course vastly larger. Their and our bodies, however, are composed of very large numbers of separate units or cells, each of which is similar in principle to the living organisms in the puddle. Each individual in the puddle consists of a single living cell and is referred to as a unicellular organism. But each of the large individual plants or animals around us owes its total individuality and nature to the coordinated activities of very large numbers of living cells; it is a multicellular organism. Living cells probably form the basis of all the life on this planet.

How, then, did this planet, which was at first presumably lifeless, come to be inhabited by these living systems? They came into existence a very long time ago, and the evidence regarding their origin is necessarily circumstantial; we depend largely on what we know about the nature of the cells that are alive at the present time. One can note that today, unicellular organisms live in water and the cells themselves consist largely of water. Water may therefore have played a major part in the creative processes that led to their emergence. It is also relevant that, early in the history of the Earth, at the time life originated, its atmosphere consisted largely of carbon dioxide, along with water vapour, a considerable quantity of nitrogen, and probably a small amount of ammonia, but with little or no oxygen. With no oxygen there would have been no ozone, and therefore nothing to reduce the amount of ultraviolet radiation in the sunlight that passed through the atmosphere and reached the surface of the land or sea.

Since 1950 there have been a number of attempts to simulate these conditions by enclosing appropriate mixtures of gases and liquid water in a flask and subjecting it to ultraviolet radiation. In other experiments electrical discharges designed to simulate the effects of lightening were used. The results were interesting. In each case, and regardless of the precise composition of the simulated atmosphere used, the water adjoining the gases was soon found to contain surprisingly large quantities of amino acids and other organic compounds, some of these being identical with substances at present found in living cells. Most of the essential building blocks of life had been formed. Energy derived from the ultraviolet radiation had presumably synthesised these materials from the simple types of molecules placed in the flask, and these had found their way into the water. If comparable processes occurred in the early history of the Earth, one can expect that tranquil bodies of surface water would have contained considerable quantities of similar molecules. The term 'primordial soup' has been applied to these postulated concentrations. We have good reason to suppose that the basic building blocks of life became widely distributed, and perhaps locally abundant, in the surface waters of the early lifeless Earth.

Yet there is still an enormous gap between a miscellaneous assortment of organic molecules, such as envisaged in this primordial soup, and the simplest type of sensitive integrated living cells existing at the present time. This is the essential gulf that needs to be bridged regarding the origin of life. The cells that are living today, using the apparatus they have inherited as a result of hundreds of millions of years of evolutionary development, have no difficulty in turning the non-living materials around them into further life. Within their own bodies simple molecules will be transformed into amino acids, which will link with others to form long chainlike molecules (peptides). These will turn in on themselves to create compact three-dimensional protein molecules. All this is done rapidly and apparently without difficulty, as though automatically. Similarly growth processes, and the production of a well-organised gene arrangement (DNA) will ensure that when a cell has grown to a certain size it will divide into two cells, each of which is an almost exact copy of the parent cell. These and many other refined processes are greatly assisted by the cell itself, which produces numerous special types of protein, known as enzymes, each of which is constructed so as to facilitate the accomplishment of one particular stage in one of these processes. These enzymes, though created by itself, are the essential agents by which it continues to grow and reproduce, and thus to create further living material of its own kind. Attempts to accomplish some of these changes without the enzymes and other apparatus that life has come to use for such purposes, and thus under the pre-biotic conditions that would have existed before such life had come into existence, have so far met with little success. We know life overcame these difficulties, but we do not at present know by which of the various possible routes this took place.

Membranes may well have played a major role. They are formed freely by rather common molecules known as lipids. These have a tail end, that tends to avoid contact with water, and a head end that has no such tendency. Hence when they are in water they align themselves side by side in two layers, with the tail end of both layers directed inwards towards one another. Collectively these lipid molecules form small hollow spheres or globules in which both the interior and the exterior surfaces are formed by tailless portions of the molecules which can be freely in contact with the water. Very large numbers of spherical globules of this type are formed.

The organic molecules that became trapped in the water within such globules would have found themselves operating in a small-scale isolated domain within which they had an opportunity to explore the possibilities of cooperating with one another; they may thus in some cases have conjointly developed procedures that made their ongoing presence in the globule more tenable. Some molecules in the world of outside water could diffuse through this lipid membrane, becoming in effect nutriment for the molecules within. Steps leading towards the eventual creation of living cells surrounded by a membrane may have been of this kind. In this context it is relevant that three kinds of molecules that would have been common in the primordial soup can mutually interact with one another in ways that lead to the creation of a molecule called pantetheine. This substance can act as an important type of enzyme. It is also used by some present-day bacteria to link appropriate amino acids into chains, thus creating peptides. This is well on the way to the formation of proteins, and some of these can act as enzymes. Developments along such lines may have helped to lead towards the eventual creation of living cells. Large and fairly tranquil pools that were well-stocked with primordial soup could have contained very large numbers of types of organic molecules and these would have freely interacted with each other, producing more types. The prospects of the evolution, by a succession of steps, of forms of proto-cells, and eventually of fully living cells, may in fact have been quite good. Plenty of time was available.

The resulting living cell would have been a very different kind of integrated system to an atom or a molecule. These latter are closed systems in which the dynamic activity is based on standing waves (p.15) which can continue to function without either intake or output of energy; it can therefore go on forever operating without change. A living cell, as also a star (p.15) is an open system, and quite different. It requires a constant input of energy to keep it in an active living state. This means that its internal condition is always continually changing. The integrative element that enables it to function as a single ongoing entity with its own individuality is based on the coordination of activities that has developed between the various types of molecule that collectively comprise the cell. These complex delicate cells are very vulnerable. Their lifespan is likely to be short, and therefore very different to that of an atom. Their continued existence depends on the rate at which they grow and then divide, so increasing their numbers, being greater than the rate at which they die from various causes. These living cells are therefore systems of a very different kind to the atoms and molecules that comprise their infrastructure. They have also developed features, for example reproduction, that are of a quite different kind to anything that had existed previously. In short, they are alive; atoms and molecules are not.

The energy that was therefore needed by proto-cells and very early living cells was presumably obtained from organic molecules that had diffused from the primordial soup through the cell's membrane, and thus entered its living domain as food. Here presumably some of the organic molecules in the food would have been transformed in ways that enabled them to be assimilated into its own living substance, so enabling it to grow; others would have been used to fuel its own dynamic processes and so make it mobile.

As this new phenomenon represented by living cells became more refined, effective and successful, their numbers would have increased until a stage was reached when the rate at which they consumed the primordial soup would have exceeded that at which fresh soup was being synthesised by ultraviolet radiation from the Sun. The position would therefore have been critical, for the further development of life would have been impossible were it not for a crucial breakthrough that enabled energy radiated by the Sun to be used in a more direct way.

There is good evidence to indicate how this dilemma was resolved. In certain places, particularly on the west coast of Australia, there are large assemblages of unicellular organisms of a kind known as cyanobacteria. These gregarious bacteria collectively surround themselves with quite bulky mineral deposits of a characteristic form, known as stromatolites. These are interesting, for their fossilised remains exist in rocks of various ages, the oldest being about 3500 million years old. Living cyanobacteria produce an organic compound known as chlorophyll which, acting as a catalyst, enables them to use the energy supplied by sunlight to convert two simple inorganic compounds, namely carbon dioxide and water, into organic compounds of a type known as sugars. This process is called photosynthesis.

The fossil evidence indicates that these cyanobacteria were already using this photosynthetic process not less than 3500 million years ago. Presumably this was the major breakthrough that appears to have been so urgently required. Those organisms that had developed and sufficiently perfected this procedure could use energy derived from sunlight to create directly within their own bodies the principal raw materials that were needed for their own living processes. They no longer had to obtain it secondhand, and their former dependence on primordial soup would thereafter have been left behind. The successful employment of this photosynthetic technique, and its subsequent wide adoption, would have had an immense effect on life's overall position. No longer would it have been limited to a fairly small number of individual organisms that were confined to those localised areas in which primordial soup was relatively plentiful. Instead these green single-celled micro-organisms would thereafter have had an opportunity to drift close to the surface anywhere throughout the vast and varied ocean waters, for wherever there was an adequate supply of sunlight, water and carbon dioxide and also some necessary inorganic salts, there they could photosynthesise, grow and multiply. The quantity of existing life, and the range of environments that it occupied, would have been increased enormously.

Photosynthesis is a complex process that takes place in a series of steps. The overall result can be expressed as $6H_2O + 6CO_2 = C_6H_{12}O_6 + 6O_2$. It therefore gives rise not only to sugar ($C_6H_{12}O_6$), but also to an incidental unwanted by-product, namely molecules of oxygen (O_2). Initially this oxygen would have passed into the surrounding water, and would have been involved in, more particularly, oxidising ferrous (iron) compounds into ferric ones. After this had been completed the subsequent oxygen would have bubbled up through the water and joined the atmosphere as oxygen gas. Here it would have been likely to oxidise any methane and ammonia, and thereafter it would have added to the amount of oxygen gas in the atmosphere. At the same time the quantity of carbon dioxide in the atmosphere was decreasing, largely due to its contributing to the formation of limestone rocks. The result was that the Earth's atmosphere came to consist principally of nitrogen, together with an oxygen component that increased steadily in quantity as photosynthesis continued through the ages. In addition a comparatively small amount of this oxygen passed into solution in the sea or in freshwater; this was very important for the fish and other life that was later to evolve in these environments.

This ability to photosynthesise seems to have ushered in a remarkably long stable period. It was an age of bacteria. Most of those equipped with chlorophyll would have been drifting in the surface waters of the seas and using the sunlight that penetrated to them to photosynthesise, so creating within themselves the basic nutriment that they needed for the business of living. The bacteria that were without chlorophyll obtained their nutriment by decomposing the corpses of dead photosynthetic ones that fell to the bottom of the sea. In doing this they were also recycling some of the materials used by the latter, so making them available again to fresh generations of photosynthetic bacteria. A very simple type of ongoing ecosystem had thus been established in the waters of the early planet Earth.

These relationships appear to have continued with little change from about 3500 million years until about 1500 million years ago, and thus for some 2000 million years. Throughout this immense period of time, which amounts to almost half the present lifetime of this planet, one can visualise the Sun continuing to radiate energy, the Earth to both circle round that Sun and to rotate on its own axis, the days and nights to alternate, the green bacteria to drift and photosynthesise and the other bacteria to decompose the available organic matter. The only major ongoing change seems to have been the steady increase in the oxygen content of the atmosphere. Looked at by hindsight, this overlong period when evolutionary change virtually ceased might seem to have been a remarkable waste of potential opportunity. Perhaps however any further major change had necessarily to await an increase in the amount of oxygen, and thus of energy, available to living organisms in their contemporary environment.

DNA–Helix

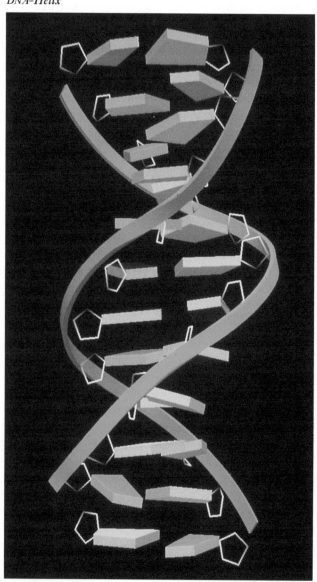

Chapter 5

REORGANIZATION AT THE UNICELLULAR LEVEL

The long period with very little change in the living world probably began to come to an end about 1500 million years ago. Fossil evidence shows that at about that time much larger types of cell came into existence. Unfortunately it tells little more, for cells are so small and delicate that their internal structure is hardly ever preserved. However it is reasonable to suppose that before long these new larger cells had much the same nature and degree of complexity as have, for instance, those unicellular organisms that, as already noted (p.30), are often present in large numbers in greenish puddles at the present time. The older smaller simpler types referred to in the last chapter, which are at the bacteria level of organization, are known as prokaryote cells, and these new larger types as eukaryote cells.

Eukaryote cells are not only larger than prokaryotes, but differ also in having a number of relatively large internal structures, known as organelles. Each type of organelle has its own special delicate and refined type of molecular organization which makes it highly efficient at undertaking one particular kind of activity. In prokaryotes, for example, the DNA (deoxyribonucleic acid) is distributed fairly widely through the cell, whereas in eukaryotes it is confined to a limited compact structure that is referred to as the cell's nucleus.

This nucleus is a very important eukaryote organelle. The DNA within it consists of long chainlike molecules. Each link in the chain carries a relatively simple side-branch, known as a nucleotide, of which there are only four types. The cell makes slightly modified copies of sectors of this DNA, and the resulting RNA (ribonucleic acid) diffuses out of the nucleus and enters one of a number of small organelles of a kind called ribosomes. Here the DNA-oriented form of the RNA provides a frame along which chains of amino acids, which are the basic components of proteins, can be readily synthesised. The order and the combinations of the four nucleotides act as a vitally important code which determines what kind of amino acids are formed, and in what sequence. An RNA molecule will therefore give rise to a linked chain of particular amino acids arranged in a particular order, both of which have been determined by the pattern of the sector of DNA that is responsible for initiating their formation. All this will in turn determine the way in which the resultant chain of amino acids will fold to form the three-dimensional structure of a protein molecule, and hence will determine in detail how this molecule is shaped and the form of its external surface. In short, then, the nature and potentialities of a cell are determined by the structure of its DNA molecules, and in particular by the way in which their nucleotides are arranged to form units or genes. Thus the DNA in the nucleus directs both the making and the functioning of its cell.

When a cell has grown large enough to divide into two its DNA strand, which lengthwise is a double structure, is precisely duplicated in its entirety. One of the two double structures formed in this way passes into each of the two new cells. Cells formed by this type of cell division, which is known as mitosis, will therefore have the same DNA arrangement as did the parent cell, and will therefore be virtually identical with it.

There are however occasions when eukaryote reproduction takes a different course. A cell will divide into two, but without the double DNA strand having previously divided lengthwise to form a new double structure. This process, known as meiosis, results in reproductive cells (gametes) that contain only half the normal quantity of DNA. Such a gamete will fuse (conjugate) with another gamete that has similarly been formed by meiosis. The single conjoint cell formed by their union will therefore once more contain a full compliment of DNA to which the two gametes, derived from two different individuals, will have each made an equal contribution. There are important additional complications, such as those associated with chromosomes and with 'crossing over', but the principal result is that the new cell resulting from the conjugation receives a set of genes that has to a considerable extent been reshuffled. It follows that new individuals created in this way will have gene arrangements, and hence resulting natures, that will differ somewhat from those of either parent. Numerous small-scale variations will therefore be produced and these will be precisely duplicated by subsequent mitotic divisions until such time as further conjugations repeat the reshuffling of the genetic material. The resulting flood of minor variations helps to keep a eukaryote species attuned to an environment that will inevitably be changing in various different ways, for at any stage there are always likely to be some individuals which happen to have inherited characteristics that make them particularly well adapted for living in their contemporary environment; these of course will be the ones that are most likely to survive and reproduce, and so to hand on at least some of these helpful genes to subsequent generations.

In addition the occurrence of occasional random changes in DNA structure, known as mutations, gives scope not merely for a reshuffling of types of genes already present, but also for the introduction of quite new types of genes, which in turn can lead to the production of living eukaryote cells with corresponding quite new characteristics.

In this context it is also worth noting that if a population becomes separated into two parts, perhaps owing to some change in the geography of an area, then the flow of genes between these two now separate populations will necessarily cease, so the total gene content, or gene pool, of each area will evolve in rather different ways. If this separation and resulting independent evolution last for a considerable period, then their respective individuals will have become so different that, if contact should subsequently be restored, then, even if they are able to mate, they will not produce fertile offspring. From this point onwards each of the two populations will necessarily proceed along its own separate evolutionary path. Two species will therefore have been created where previously there had been only one. This process is known as speciation.

The scope for the creation of numerous fundamentally different categories or species of organisms, and for much small-scale diversity within each category, was therefore greatly increased with the coming of the eukaryote type of cell. Clearly the influence of DNA is crucial; on the one hand it is merely a complex, but lifeless, organization, whereas on the other its own evolving structure is both guided and constrained by the natures of the living organisms that it helps to create, since patterns of DNA that do not organise effective cells will soon disappear along with the cells for which they have been responsible. In prokaryotes (i.e. bacteria) there is an occasional exchange of genes between individuals, but the process is not systematic or highly organised. The coming of eukaryote cells that were excellently geared to a production of a stream of heritable variations meant that an inbuilt system of automatically creating much biodiversity had been established. The evolution of life could thereafter get fully into its stride.

In prokaryotes the chlorophyll that promotes photosynthesis is also widely distributed throughout the cell, whereas in eukaryotes it is confined to a compact, clearly demarcated organelle known as a chloroplast. Thus if the nucleus, through its DNA, acts as a cell's organizing agent, then the chloroplast, through its chlorophyll, acts as its powerhouse. The microstructure of this organelle has the form of numerous minute parallel sheets each supporting a closely-packed layer of chlorophyll molecules; these are therefore excellently exposed to sunlight. These chlorophyll molecules apparently provide a framework that presents the water, carbon dioxide and sunlight to one another in a way that enables the energy associated with the sunlight to dissociate the water molecules into their hydrogen and oxygen components; the resulting hydrogen becomes linked with the carbon-oxygen complex provided by carbon dioxide; this unstable unit is then converted, by a series of fairly simple steps, into the stable carbohydrate called sugar. The chlorophyll merely acts as a framework or catalyst that facilitates the initial transformation; its molecules remain unchanged, and the process can therefore be repeated again and again, using fresh raw materials on each occasion. This important new form of creative organization can thus facilitate the crucial initial step in the manufacture of a steady supply of sugar for as long as the cell requires it, provided always that sufficient carbon dioxide, water and sunlight are available. Most of the oxygen derived from the disintegration of water is not needed and is excreted as bubbles of oxygen.

The sugar formed by photosynthesis diffuses through the cell, and some of it reaches very much smaller organelles known as mitochondria. Here the manufacture is such as to facilitate a respiratory process which is the reverse of photosynthesis; oxygen combines with this sugar which is thereby broken down, again by a series of steps, the final product being carbon dioxide and water. Most of these steps involve an output of a small amount of energy, and this energy can be used in the mitochondrion to convert a molecule known as ADP into a slightly different one, ATP. These ATP molecules reach positions in the cell where energy is required, perhaps for chemical synthesis, or for rendering the unicellular organism mobile. The energy that was transferred to a mitochondrion can thus be made available at a site where it is required merely by transforming ATP back into ADP; the latter then passes back to a mitochondrion and is there recharged. The overall result is that radiant energy emitted as light from the Sun is transformed into chemical energy by the photosynthetic process, and this is subsequently made available through the medium of ATP molecules; each of the latter acts like a minute mobile battery that delivers a very small and precise quantity of energy to a precise location at which the cell requires to bring about some precise small-scale molecular transformation. Precision, based on very small-scale physical events, is thus a crucial feature of the infrastructure of a eukaryote type of living cell.

Molecular changes that take place step by step are characteristic of eukaryote living cells. Each step is usually associated with an enzyme whose surface shape precisely fits that of the two molecules that are required to interact with each other. This enables them to be brought together very closely and precisely, and they will consequently undergo the chemical change in question exceedingly rapidly. One is involved here with time intervals of picoseconds (each of which is a millionth of a millionth of a second) and with space separations of nanometres (each being a hundred thousandth of a millionth of a metre).

Thus the detailed form of the external surface of an enzyme is crucial because it needs to be complementary to, and so to fit, that of the molecules it is to accommodate. Usually, therefore, any given enzyme will promote only one type of chemical change. Because enzymes are so specific they provide a means by which a cell can establish feedback mechanisms. When the substance that is being produced reaches a concentration that is sufficient for the cell's immediate needs, then the substance will react with the relevant enzyme in a way that renders the latter temporarily inoperative. The production of the substance is therefore suspended, but it can be resumed again when usage has reduced its concentration. Such feedback mechanisms prevent the cell being flooded by the substance. They thus provide a means by which a cell can regulate, and so also integrate, its own internal activities. Thus in eukaryote cells precision, efficiency, rapid production and self-regulation can work closely together.

The need for self-regulation is far ranging. Any individual organism, if it is to survive and grow, must be able to put itself into positions where it can obtain the energy and the raw materials that it needs and also avoid the hazards that beset it. It must therefore be capable of responding to a whole range of contingencies that are continually changing. To do this it must be sensitive to its surroundings; also all its organelles must be capable of working together in a way that enables its total living system to respond in an effective manner to what is sensed. There will, for instance, be changes in the intensity of the light received, varying from bright sunlight to cloud cover and to the dark of night, and also differences in the temperature of the water, and in the kind of objects that it touches; it must be sensitive and responsive to many such situations. The buildup of substances within itself needed to meet these various contingencies will also vary. The effect of such variables will extend throughout its whole cell complex. It will therefore be in a continual state of flux, and in this sense also it will be very much alive. The fact that its internal organization is based on step-by-step molecular transformations, that each such step usually requires the mediation of an enzyme, and that recycling and feedback mechanisms are often involved, together give to a cell great flexibility. Its functioning can be temporarily deviated in many different directions depending on what contingencies arise, yet throughout it will behave as a single integrated unit and it will tend always to return to approximately its former state.

These few comments may suffice to indicate that the relatively new science of molecular biology, greatly aided by instruments based on new developments in physics, has demonstrated that eukaryote cells are able to make use of exceedingly minute quantities of energy in ways that are both elegant and precise. The molecules that they synthesise comprise not only the living material of the cell, but also the molecular apparatus that makes such production possible. It could be likened to a workshop or factory in which not only are goods produced, but also the necessary machine tools are made, kept repaired and multiplied. Since the goods are largely converted into more tools, so promoting further growth, these two aspects of a cell's activities are closely intertwined. Furthermore, when a factory has reached a certain critical size it employs mitosis to divide into two separate similar ones. Also on occasion, and even more dramatic, halves of two separate factories amalgamate or fuse, as a result of a conjugation, and in doing so they remodel the finer details of their management apparatus into a new form. Minor diversities in factory types therefore result. The factories currently engineered and managed by us humans seem by comparison simple, crude and lacking in refinement.

Clearly the revolution that resulted in the creation of eukaryote cells from earlier prokaryote ones was exceedingly important. Direct evidence about how or why this occurred is lacking. Instead we have to rely on circumstantial evidence, and conclusions are necessarily controversial. For many years Lynn Margulis has maintained that a eukaryote cell originated as a result of a nonphotosynthetic bacterium that would therefore have normally fed on organic food engulfing a photosynthetic one, but that the latter, instead of being digested and so used as food, continued to live there by establishing a working relation with what had by then become its host. It thus became a chloroplast which used its chlorophyll to synthesise sufficient sugar for both the parties concerned; in return it was relieved of the responsibilities of living an independent life. Both therefore benefited from this symbiotic relationship. Such combined cell-units would have been larger and probably more efficient than single cells, and might therefore have provided a favourable environment for the evolution of specialised organelles within themselves. This suggestion regarding the origin of eukaryote cells seems now to have gained wide acceptance.

Another major factor promoting fundamental change may have been the ready availability of oxygen that had resulted from hundreds of millions of years of photosynthesis by thousands of millions of prokaryote cells. This made aerobic respiratory processes possible, for instance in mitochondria, and this in turn made far more energy available for various unicellular activities. Some eukaryotes seem to have used this to give them greater mobility and dynamism, which in turn enabled them to hunt down, consume and digest other unicellular organisms. This would have provided them with a first class source of nutriment at little cost. Here the well-known unicellular eukaryote *Amoeba* comes to mind. Such organisms could then afford to dispense with chloroplasts and photosynthesis, for they could merely steal the nutriment that they needed by killing and consuming other organisms that had already done the work of synthesising it. A whole new world of ecological relationships—the animal and the plant—the predator and the prey—was thus being brought into existence.

In short, the emergence of the eukaryote level of cell organization was probably the result of the fusion of two cells that were at the preceding prokaryote level. This fusion gave scope for the evolution of organelles within the larger conjoint cell, and this internal specialization in turn gave rise to important new potentialities, one of these being the subsequent evolution of multicellularity. The long prokaryote static period had been brought to an end. The emergence of this eukaryote type of cell organization was an exceedingly important event in the history of life on the Earth.

Fossil Trilobite, Wales UK

Jeff Spielman, The Image Bank

Chapter 6

MULTICELLULARITY

The last two chapters have been concerned with a period when the life on the Earth consisted merely of microscopic single-celled organisms that lived in a watery medium. At the present time, by contrast, many relatively large animals live in this same medium; there are, for instance, fishes, prawns, mussels and sea anemones. In each case the body of these creatures consists of very many, indeed often of many millions, of separate living cells, each with a eukaryote type of organization. These many cells cooperate with one another so effectively that conjointly they give rise to a single well-integrated multicellular individual.

The method of reproduction of these multicellular animals throws light on the general situation. Take salmon, for example. There are two types, namely males and females. Adults of both sexes make their way up from the sea to near the source of a river. There, in some suitable pool, a male and a female, when in close proximity to one another, simultaneously discharge their reproductive products into the water. In both cases this substance consists principally of single cells or gametes which, as in unicellular eukaryotes, have recently been formed by meiosis (p.38); each gamete has only half the full compliment of genes. In the case of the female fish these reproductive cells are large and immobile, being heavy because they contain nutritive material that later will be needed by the developing embryo. The male gametes, or sperm, are small light cells that are equipped with a flagellum (p.29) that enables them to swim through the water and so reach and fertilise the egg-cells extruded by the female partner. Some form of courtship or sexual stimulation will be needed to ensure that the potential parents discharge their reproductive cells at the same time and place. The differentiation of adults into two sexes, and the sexuality associated with this, has arisen as a device that ensures that the relevant gametes will be able to make contact with one each other and that sufficient nutriment will be available for the embryo.

Conjugation restores the full compliment of DNA to the fertilised ovum. It also reshuffles the genes and so provides a mechanism that automatically gives rise to a variety of rather different types of offspring. In unicellular eukaryotes the cell formed by a conjugation undergoes a series of mitotic divisions (p.38); the resultant cells then swim away as separate unicellular individuals. In multicellular organisms, such as salmon, there is a similar series of mitotic divisions but the resulting cells remain together in close association within a common coating, and by their cooperative activities these cells conjointly give rise to a well-integrated multicellular embryo that will grow into a new adult. This cooperation was probably facilitated because, in the course of their earlier history, the eukaryote cells that pioneered this revolutionary behaviour had become sensitive to the detailed stimuli they received from their environment and would have become well able to respond to what they sensed. In the new setting their environment would now include the presence and the activities of other cells essentially similar to themselves. It was to this new situation, as well as to the outer world beyond, that their inherited ability to sense and to respond would now be directed.

One consequence of this type of origin has been that the rhythm of life of a multicellular species undergoes remarkable fluctuations from a multicellular condition back to a unicellular one, gradually building up again to full multicellularity. One major advantage of this rhythm was that it enabled multicellular species to reshuffle genes and assimilate mutations, and hence develop a wide range of biodiversity (p.39), that had already been developed by their unicellular eukaryote ancestors. Linked with this were the very important advantages that can accrue from increased size and complexity. The scope for the evolution of a great diversity of living things was enormously enhanced by the coming of multicellularity. The qualities inherent in these emergent forms of life were continually being pruned, refined and rendered more efficient by the processes of 'natural selection', which were always related to the particular way of life the species in question was adopting. Salmon, with its great internal complexity, its overall integration, its abilities and its beauty, is a longterm product of processes of this kind.

This periodic return to a unicellular condition also gives rise to new complications. Each new individual of each new generation begins its life at conception as a single cell. It has then to climb the long and difficult path leading from unicellularity to its own particular type of multicellularity. In its early evolutionary history, its level of efficiency will have been low, but so will have been the level of competition that it encountered. Improvement will have been a gradual process involving numerous successive steps in which each constructive accomplishment provided a platform for potential further new developments. Improvement involved two aspects. On the evolutionary side, each new departure has to come into existence, has been assimilated into the living organization, and has proven its worth in the ubiquitous selective testing. On the embryological side, each new reversal to the unicellular condition requires that each new departure that has proven helpful for survival be recreated during the development of each new individual. These two aspects are interrelated. The DNA, the RNA, the amino acids and the proteins, as also the mitoses, the meioses and the conjugations, all play a part. The salmon provides an example of the kind of results that are possible.

The actual problems that needed to be overcome on the road to the creation of a salmon have been formidable. The successive mitotic divisions that follow fertilization would, if left unguided, give rise merely to a spherical mass of more or less similar cells. Our salmon embryo is more, however, than just a ball of cells. Being an animal it will have to be able to sense its environment and move about within it, and thus be able to find, catch and make use of the food it will need as soon as the nutriment originally bequeathed to it by its fertilised ovum has become exhausted. A ball of undifferentiated cells would be very ineffective in this situation. In an embryo a group of genes, sometimes referred to as master genes, comes into action at a very early stage and induces the cells to become organised to form a bilateral structure with front and rear ends, and to initiate in appropriate positions such crucial organs as the embryo's mouth, eyes, brain and spinal cord. It is this early essential structuring, organised by these master genes, that determines what general type of animal will emerge. In the case of the salmon, the creature that hatches from its egg, though very small and vulnerable, will nevertheless be able to locate, reach, consume and digest the food it requires.

Secondly, it is important that the cells that form the crucial elements in any organ, such as for example the liver or a muscle, are not just typical cells, for they have to become specialised in ways that enable them to contribute very effectively to the functions of the organ of which they become a part. Such cells may in fact be very different from those in their initial state as regards their appearance, shape, structure and mode of functioning. It is the collective activities of specialised cells that make an organ capable of discharging its own particular function with such remarkable efficiency. The change from an undifferentiated cell to a specialised one seems to be brought about approximately as follows. All the cells in an embryo have the same series of genes. The region in an embryo where a particular organ is to be initiated exerts some kind of influence, possibly of a chemical nature, which activates a particular gene in those undifferentiated cells that are close by. This gene, when thus activated, will cause the cell to transform itself into the type that will be required by the organ in question. Here time, place, needs and actions are simply and neatly brought together and coordinated.

A third problem arises from the relatively massive bulk of cells of any moderate-sized multicellular animal. Its overall activities (i.e. its metabolism) must operate in a way that always maintains within its body an internal environment that suits all its constituent cells; this enables them to live active healthy lives and thus to contribute fully to the health of the total organism of which they are a part. These cells have urgent needs that must be constantly met; for instance, an adequate supply of nutriment, and also of oxygen, and their excretory products must be removed. In salmon, as also in all other vertebrates, this is accomplished by means of a circulatory system in which blood is continually pumped round the body by rhythmic muscular contractions of a heart. Good health is a two-sided process; a healthy body maintains healthy cells, and the healthy cells maintain a healthy body. At all times the whole and the parts are mutually influencing one another.

A fourth point is that if a multicellular organism is to survive and prosper the numerous and varied activities that take place within its body have to be coordinated. It has to become a fully integrated living system. The circulatory system has a specific function here, too. There are special organs (called endocrine glands) which secrete types of molecules known as hormones into the blood stream. These hormones are transported far and wide, and so reach and affect the activities of cells in other parts of the body that are adapted to respond to them. They act as chemical messengers that help to keep an organism's metabolism balanced.

This method of communication and coordination, which is rather slow and imprecise, is supplemented by the presence of quite different specialised cells, known as nerve cells, that collectively comprise a nervous system. A nerve cell may, for example, have a long extension, its axon, which is in close contact with some sensory cells, for instance those in the retina of an eye. This nerve cell can then readily convey information, in the form of electrical impulses, from these sensory cells to the main body of the nerve in question, and these are then conveyed onwards through its delicate branched extensions to further nerve cells. In this way information about the environment around an animal, can be rapidly conveyed to the concentrated mass of nerve cells that comprises its brain. This brain also receives information about its body's internal condition. Networks of nerve-cell communication in the brain can collate such information, determine an appropriate response, and put this into effect through the medium of so-called motor nerves which stimulate the appropriate muscles to contract in an appropriate manner. All this can take place fairly rapidly and with precision, in contrast to the hormonal method of coordination, which is slower and less precise and serves a different purpose.

The individuals most likely to survive are those with genes that cause them to have a strong urge to survive. These genes, and hence also this urge, will therefore be inherited. This continuing selective pressure, acting through innumerable generations, has led this urge for self-survival to become strongly embedded in all multicellular organisms, whether they be salmon or humans. Similarly it is those individuals that have a strong sexual urge that have been the most likely to reproduce; this urge has therefore become another important and deep-seated feature of the present planetary scene. Past and present are closely intertwined.

It may seem almost incredible that problems of this kind was resolved so effectively, that such highly complex living machinery were created, and that all this became well integrated and also well adapted to the habitat in which the relevant species live. Yet the existence of the species now existing, in our case example the salmon, shows that all this was in fact accomplished. Of course, like Rome, this evolutionary construction was not built in a day; it has taken some 500 million years for successive generations, each with their individuals struggling to survive and reproduce, to bring multicellular organisms to their present very high level of perfection.

A number of very different types of multicellular organization exist in which some representatives have evolved a very high level of complexity and refinement. Examples are salmon, lobsters and squids, belonging respectively to the groups Vertebrata, Crustacea and Mollusca. In their progress towards their respective present living natures they have confronted similar problems regarding a need for effective respiratory, excretory, circulatory, nervous, skeletal, locomotory and other systems. They have each resolved these problems by evolving organs of quite different kinds, though with comparable functions. Even their overall architecture is very different. Each have become complex, highly integrated, multicellular organisms, but this has been attained by progressing along very different evolutionary pathways. To attain such ordered structural and functional complexity would, as already noted, seem a formidable achievement. In fact, however, this type of overall evolutionary development has taken place largely independently in a number of very different groups of animals. Clearly it is well within the normal longterm creative capacity of eukaryote cells that are employing a multicellular mode of life.

When multicellular organisms became relatively large, their bodies needed skeletal support and their muscles needed firm anchorages. This required the evolution of appropriate hard structures, such as bone or chitin or shell. Hard structures tend to resist decay after death and so are able to be preserved as fossils in rocks that are eventually formed from the layers of silt that accumulated on the sea floor. From the beginning of the Cambrian period onwards, and thus for about the last 600 million years, the preservation of hard parts in the form of fossils clearly shows that certain kinds of organisms were alive at certain periods in the past. To research the past history of life from this point on, we no longer rely almost entirely on evidence provided by organisms living at the present time. Furthermore, thanks to recent new techniques associated with rates of radioactive decay, it has become possible to determine the age of some of these fossil-bearing rocks not merely in relative terms, but as fairly precise estimates of the actual number of years that have elapsed since the organisms represented by the fossils were alive. This geological record is of course far from perfect. It depends, among other things, on whether a local situation did in fact result in fossil being formed and, if so, on whether the rocks that conserved this evidence happen to be still extant and accessible for study. Nevertheless, in spite of imperfections, from the Cambrian onwards, much direct evidence does exist regarding the past history of life on planet Earth.

It happened that at a fairly late stage in the Cambrian, more than 500 million years ago, there were sometimes local circumstances that prevented the decomposition of some of the soft parts of animals; hence these parts also became fossilised, and now provide important additional information. In one instance the rock that had originally formed from sediments that had accumulated on the seabed, has subsequently been raised by mountain-building, and is now located high up in the Rocky Mountains in British Columbia. Study of the fossils in this rock, which is known as the Burgess Shale, shows how limited the information that is provided by the presence of hard parts alone is. The additional evidence here demonstrates, for example, the presence of two new groups of worms, of several types of primitive crustaceans, of some sponges, and of a fossil that is believed to be a proto-chordate and thus an early forerunner of the vertebrate type of organization.

The unicellular photosynthetic organisms drifting as plankton near the surface of the sea were then—and still are—the ultimate producers of virtually all the nutriment that existed in the sea. This plankton would have provided an almost inexhaustible source of nutriment for any small multicellular animals that became capable of harvesting it. They, like the unicellular Amoeba (p.43), would then have been adopting an animal mode of life. Such a rich nutritional source, once effectively tapped, would enable the creatures in question to evolve into larger and more diversified multicellular organisms. Some of these adopted a more economic procedure, namely that of preying directly on the consumers of the plankton, instead of on the plankton itself. A complex food network involving a dynamic system of ecology based largely on predation would thus have been created. The fauna of the Burgess Shale was probably at this stage, and represents the situation at a time when multicellular evolution was just beginning to get into its stride. Quite a number of groups of organisms existing at that time subsequently became extinct; this also suggests that this shale is providing information about an early stage in this type of evolution, when competition was not as yet fierce, and when numerous possible ways of living could be tried freely. Some succeeded, but there were also many that soon fell by the wayside.

The longterm success stories as seen in the sea today, some 500 million years later, include a number of major multicellular groups. First there are the coelenterates; these are rather simple types of multicellular organisms; well-known examples are jelly fish, sea anemones and corals. Secondly there are various kinds of worms, some of them swimming freely and some living within hard tubes formed by themselves from which, when it seems safe, they partially emerge to gather food by filter feeding; when doing so they display great beauty. Thirdly, there are echinoderms with their radial symmetry and the tube-feet that assist their movements; familiar examples are star fishes and sea urchins. Fourthly there are mollusca; many of these have calcareous shells; familiar types are mussels and octopuses and various types of snail-like creatures. Fifthly there are crustaceans with their numerous and varied types of appendages and their chitinous external skeletons; they vary in size from tiny copepods to such relatively large animals as crabs and lobsters. Lastly there are vertebrates, represented in the sea mainly by fishes. Each of these six major success groups have developed much diversity within themselves, and hence numerous subgroups, some with many species, and some of these are represented by great numbers of individuals.

There have, of course, been numerous major setbacks and multiple extinctions during this long period, but broadly-speaking multicellularity has been prospering from the Cambrian onwards. As the quantity, quality and diversity of this life in the sea increased, and the various available ways of life became increasingly filled, so too did the need for each species to find and exploit effectively an appropriate niche within the complex relationships that were resulting from the mutual interactions between the various species operating there. The structure and mode of life of each species had to become adapted to meet this end. Some species evolved in this manner, becoming established members of an ecological complex, continuing to adapt as further contingencies arose. Such species have been able to continue in existence, and in some cases to flourish. Species that failed to adapt waned and became extinct; the opportunities thus left available were taken by other species that became adapted according to the circumstances then prevailing.

As in the past, the sea today offers four main types of habitat within which such ecological dramas can take place. Firstly, and biologically the most important, is its surface layer. Bounded by the atmosphere externally, sunlight penetrates this zone, reaching into it with decreasing intensity down to a depth of a hundred or fifty metres. In this uppermost zone planktonic unicellular organisms are abundant, and it is here, and here alone, that photosynthesis takes place. It is therefore also here that almost all the organic nutriment that supports all the life in the sea is initially generated. In this zone there are many small multicellular animals that feed on this plankton, usually by filtering these minute organisms from currents of sea water that they pass through their filtering organs. There are also larger animals in this zone, such as small fishes, that prey on the organisms that live by filter feeding. These are themselves eaten by smaller numbers of larger animals, and so forth, until at the top of the pyramid there are a very small number of adults of very large species that prey on smaller ones and are themselves not preyed upon by anything. An adult male and female cod, which are near the apex of the pyramid, will produce about six million fertile eggs. The tiny fish that hatch will take their place drifting with the plankton, eating it and being themselves eaten. On average only two individuals, out of the original six million, will survive long enough to make their way upwards towards the apex of the pyramid and then, in maturity, produce a further six million fertile eggs. The rest will perish earlier.

The portion of the sea that lies below this productive uppermost zone comprises three main regions. The first is the furthest distant from the continents. Here, well beyond the continental shelves, below the top zone, an extent of water lies some three or four miles deep that supports comparatively little life. Below this there is the ocean floor. The water near to this floor is at too great a depth to be stirred by winds or currents, and is therefore very tranquil; it is also very cold, poor in oxygen, and totally dark. Corpses of organisms from the productive zone above fall gradually down to the floor, providing some organic matter. Most of this is decomposed by bacteria, and it also provides some nutriment for the few highly specialised multicellular animals that struggle to survive in this difficult benthic desert-like environment.

However in certain regions of this deep ocean floor there also exist the rift-valley type of systems that are formed where tectonic plates are moving away from one another (p.26). About twenty years ago it was discovered that there are places within these rift valleys where hot water containing hydrogen sulphide, sulphur and metal-bearing minerals gushes upwards out of the mantle below into the cold sea water above. These hypothermal vents were found to be local warm oases that supported abundant life on this desert-like ocean floor. It seems that under these conditions bacteria of a special kind, known as archaebacteria, are able to change the hydrogen sulphide molecules into hydrogen and sulphur, and to use energy produced by this chemical reaction to synthesise the organic compounds required for their own metabolism and growth. This process appears to be comparable with photosynthesis, though of course different from it, and the two modes of synthesis have doubtless been evolved quite separately. A remarkably dense and varied assemblage of multicellular animals is also associated with these vents, being nourished, directly or indirectly, by the archaebacteria. These animals include tube-worms, mussels, limpets and small fishes, many of them specialised and different from species existing elsewhere. A detailed study of these strange and unexpected communities may prove highly relevant to our understanding of the early evolution of life on this planet.

Nearer to the margins of the oceans the underwater extensions of the continents ensures that the sea is relatively shallow. Here the water near the bottom is usually fairly warm and well oxygenated, the sediments contain a considerable amount of organic matter, and currents and sometimes storms stir the water and tend to mix and distribute both organic and inorganic materials. Also the depth of water separating the top productive zone from the bottom one is small or fairly small, and there is likely to be some traffic between them. The sediments on the sea floor and the adjacent overlying water therefore often teem with life. It is these regions that support the concentrations of fishes that now attract the attention of the competing, highly mechanised fishing fleets of the world.

Lastly we reach the very shallow water near to coasts. Here the top and bottom productive zones meet one another and merge. This region is therefore potentially very productive. Here, however, the situation is complicated by such special factors as tides and occasional storms, the large quantities of fresh water and mud poured into the sea by rivers, coasts that may be rocky is some places and sandy in others and now also by pollution and disturbance caused by great concentrations of land-based humans seeking relaxation by the seaside.

Chapter 7

LIFE ESTABLISHES ITSELF ON LAND

While life, both unicellular and multicellular, was flourishing in the water of the seas of planet Earth, for instance at the time the Burgess Shale was formed (p.50), the continents remained merely barren zones that protruded above the seas. Their land surfaces presented a difficult problem for organisms that had presumably originated in water and had always subsequently lived in that environment. On a land surface there was no congenial all-pervading watery medium; indeed virtually the only water present was that derived from rain. As far as we know the land surfaces of the Earth remained a wilderness devoid of life, somewhat like the surface of the Moon today, until about 420 million years ago, and thus long after the seas of the world had become extensively populated. The fresh water had by this time also become populated by abundant unicellular life, and various early types of fishes had moved from the sea into rivers and lakes. Fresh water, as well as salt, therefore contained plenty of life, though here the degree of diversity was much less.

Sheltered backwaters where land and fresh water meet in tranquility may have been particularly helpful in bridging the gap between water and land, for here there would have been places where there was neither water nor land as such, but instead an intimate blending of the two. One can visualise that here green unicellular eukaryote organisms, such as those that are often so numerous in puddles today (p.29), could have settled on the damp muddy surfaces. Mud would have been present below them, water around them, and sunlight coming to them from above. This combination would have provided excellent opportunities for photosynthesis, and therefore for growth followed by mitotic cell divisions, so giving rise to a mat of cells that, lying on the surface of the mud, would have been in immediate contact with one another. Because such cells were already responsive to their environment they would, in principle, also have been capable of responding to one another. They could, by their conjoint activities, have created simple integrated multicellular plant organisms. The tranquility, and the conjunction of water, sunlight and mud, may therefore have favoured the development in such places of some initial form of plant multicellularity.

Much further evolution would of course have been needed .before plants could establish themselves on the surface of drier land in the vicinity of the water's muddy edge. There are rocks that contain well-preserved fossil remains of early land plants. They were multicellular organisms of a rather simple type. The part situated below ground consisted principally of cells that were presumably well able to absorb water and inorganic substances that would have been present in the ground surrounding them. This rootlike portion also provided an anchorage. The remainder of the plant had the form of slender stems that extended into the air above the ground. Many of the cells in these stems contained chloroplasts equipped with chlorophyll, enabling them to use carbon dioxide derived from the air and water drawn from the ground by the 'roots' to manufacture sugar by photosynthesis (p.34). The stems were covered by a thin, but tough, cuticle which limited evaporation and so prevented these early plants from becoming desiccated. There were numerous small openings known as stomata in this cuticle which allowed carbon dioxide in the atmosphere to reach the chloroplasts. Opening or closing of these stomata could be regulated in accordance with the plant's carbon dioxide requirements at any given time. It was of course necessary, as also in animals, for the cells that collectively comprised these multicellular plant organizations to be able to coordinate their activities in a way that maintained an internal environment within the interior of the plant that suited the cells living there. The whole living plant system could thus grow.

Reproduction in this new setting raised its own special problems. Each plant was anchored to one spot; it was this that enabled it to obtain the water that was so vital. Yet dispersal was essential. So was the maintenance of genetic diversity. These needs were met, firstly, by the production within special capsules of very large numbers of small light cells known as spores; these resulted from a series of mitotic cell divisions, and therefore each contained with its nucleus a precise copy of the DNA that was present in all the cells of the parent plant. When the spores were ripe the capsule that housed them suddenly burst open, and the light dry spores were then carried far and wide by any slight breeze that happened to be blowing. Radiant energy that gave rise to heat, and consequently to wind, was thus being used for their random dispersal. The eventual result of this evolutionary development was that spores that landed in some suitable location germinated there, forming very small rudimentary plants each of which produced either male or female gametes. The male gametes, equipped with flagella, swam through the thin film of water that was always likely to be present in any damp situation, and so reached and fertilised an adjacent female gamete that had been produced in a comparable way by some other rudimentary plant. The egg cell resulting from this conjugation would then, by successive mitotic cell divisions, set about creating a new multicellular plant of the appropriate type. Thus by alternating processes, the one involving nonsexual spore formation and the other sexual conjugation, the plants that were pioneering their establishment on land were able to provide adequately both for their wide dispersal and for the maintenance of their diversity. Both were essential for the longterm success of this evolutionary experiment.

Success soon gave rise to further problems. Sites where such plants could flourish would have been limited, and there would have been competition for space and light. Multiple branching of stems, with each branch supporting numerous flattened fronds, as in ferns, and with each frond bearing numerous chloroplasts that were exposed to sunlight, would have increased efficiency. As regards sunlight, so crucial for photosynthesis, the advantage would have gone to the taller and the disadvantage to the less tall. The struggle for light, and therefore for height, seems to have gone on apace; initially typical plants had slender up-growing stems only a few centimetres long, but before the end of the Devonian, some 50 million years later, there were quite large fernlike plants. Their stems were strengthened by wood, a strong yet yielding substance that provided adequate support and permitted an elaboration of form. Later, and still in damp low-lying areas, dense Carboniferous forests grew, with their huge lycopodium, horsetail and fernlike trees. All through this gradual transformation any plant species that failed to keep up with the race for height would have either been squeezed out of existence and become extinct, or would have become adapted for life in some special ecological niche.

In these forests there would have been a local recycling of materials. Substances drawn up from below ground, or drawn in from the surrounding atmosphere, would have been used by the trees for their living and growing. Presently, after these trees had died and fallen to the ground, they would have been decomposed by bacteria on the swampy forest floor; this made the essential raw materials that they had used available once more for fresh generations of trees. Also the land surface, initially infertile, would have become rich in organic matter, and therefore fertile. At certain times and places the bacterial decomposition remained very incomplete, probably because the activities of the microbes was inhibited by lack of oxygen within the water-logged debris that accumulated. It is the fossilised remains of these partially decomposed plants that has now become coal; latent within it is still the stock of chemical energy that these trees created within themselves as a result of photosynthesis so many millions of years ago, and for this reason it has come to play a very important part in our present human activities.

The damp low-lying portions of the land surface of the Earth became covered with an increasingly thick green mantle during the Devonian and Carboniferous periods. This became a shaded, cool and moist environment that provided green plants and plant debris on which animals could feed. The plants, by their presence, would have been transforming low-lying portions of the land from hot dry barren places to cool, shaded luxuriant ones that came to teem with life.

Plant life in the water (apart from seaweeds) had remained essentially unicellular. The land plants had to pioneer their own multicellularity, and of a type appropriate to the new environment in which they were becoming established. In the water there were already numerous types of multicellular animals, some of them quite complex, and each closely adapted in both their structure and function to some particular ecological niche in their aquatic environment. Animals differed from plants in that they did not have to fashion their type of multicellularity in accordance with the nature of this new land habitat. To those multicellular animals that happened to be able to make the difficult transition from water to land, a whole new world of future evolutionary possibilities lay open.

Arthropods were probably first on the scene. There were already millipede-like creatures were present in the late Silurian. As trees grew taller an ability to fly would have been useful, and in the early Carboniferous there is evidence of numerous types of insects, including very large dragonflies whose fossil remains can be spectacular. It is only in the last stage of growth, when insects are mature, that their wings become functional; their principal use was therefore probably in finding mates and in facilitating the dispersal of their offspring. Some members of another subgroup of arthropods, namely arachnids, had also became well adapted to life on land, as is demonstrated by the fossil remains of large scorpions, and also of spiders, in rocks formed during the Carboniferous period. It is clear that at this time there were not only animals that were obtaining the nutriment they needed by feeding on plants, but also others that obtained it less directly, namely by preying on these plant-eating animals. Dragonflies, scorpions and spiders would all have been in this latter category. Evidently, an ecological network of food relationships that was comparable with, though different from, the networks that had been elaborated earlier by organisms that had always been living in water, now existed on land.

Turning to vertebrates, the fossil record indicates that members of one of the major groups of fishes made the important transition from an aquatic to a terrestrial habitat, thereby becoming amphibians. Perhaps they did not make this transition voluntarily; it may initially have been forced upon them when pools in which they had been living dried up, so leaving them stranded on a wet land surface. Adjacent plant cover could have provided shade and a moist cool atmosphere, and invertebrate food would have been available. At first they would have been very incompetent on land. They could only crawl or wriggle their way forward, not walk, for limbs that had only recently been evolved from fins were for a long time unable to lift their bodies off the ground. They could not venture for long into regions where the air was not kept cool and moist by a mantle of vegetation, for they had no external covering comparable with the cuticle of plants, and would soon have become dehydrated. In addition, they would still have been dependent on a source of water that would have enabled a male's sperm to swim to reach a female's eggs, and in which the resulting offspring could live the early stages of their lives. These limitations would have been somewhat compensated by there being as yet no other vertebrates with which they had to compete.

In some such ways new types of plants, along with certain animals that followed them, established bridgeheads on the low-lying margins of the land masses of the world, so forming the Carboniferous forests. For living organisms to extend their range from here onto the drier higher hinterland involved another set of problems that were also difficult to resolve. As always, it was plants that had to be the initial pioneers, for it was they that provided the necessary food, shade and local environments that enabled animals to subsequently follow. This chapter ends with some brief comments on the spread and subsequent evolutionary development of plants, and of the insects that are closely associated with them, in these more difficult environments. The following chapter then gives a similar consideration to vertebrates.

In these drier environments plants had to evolve a very effective waterproof covering that could help them to conserve the water within their living substance. They needed spreading roots that would be able to absorb sufficient of the rather small quantity of water that would have been present in the ground where they were living. Also it was necessary for these plants to evolve a method by which male gametes could reach their female counterparts in the absence of water. The previous method of spore-formation (p.56) was further elaborated by the production of spores of two types, one being much larger than the other. The larger ones (megaspores) remained closely associated with the parent plant and, by germinating there, gave rise to a minute rudimentary next generation plant that produced female gametes. Female gametes thus remained functionally more or less a part of the initial primary plant. The small spores (microspores) became detached from the parent plant, and they were produced in such vast numbers that some of them were likely to be blown to a position where female gametes were being formed on another plant. They would germinate there, so giving rise to even more rudimentary next generation plants that produced male gametes. Some of these would be located so very close to female gametes that they could reach and conjugate with them without any need for external water. This proved to be a very successful strategy. Already some 300 million years ago plants of a type known as gymnosperms, which include modern conifers, had evolved this type of lifecycle. They produce vast quantities of microspores—in this context better known as pollen grains—and rely on the vagaries of the wind to convey some minute, yet nevertheless sufficient, quantity to the required position on another tree.

Insects responded to the challenge arising from living in dry regions by evolving two distinctive phases within their own life histories. In their earlier or larval stage their structure and manner of behaviour are adapted to feeding more or less continuously on some type of food, perhaps the leaves of some particular species of plant. It thus concentrates on growing as fast as possible, so reducing the length of time spent in this vulnerable stage in its life. A quiescent or pupal stage follows, during which most of its anatomy is comprehensively reconstructed. The outcome of this metamorphosis is the emergence of an adult insect that is not only equipped with wings (p.58), but now also differs in almost all respects from its former larval state. During this adult phase it does not grow, but is concerned with mating—which does not require the presence of water—and, in the case of the females, with flying in search of suitable places in which to lay the eggs that will give birth to a new generation. Insects with life histories of this type include butterflies and moths (Lepidoptera), bees, wasps and ants (Hymenoptera), and flies (Diptera).

The life history of these insects is of special interest because it came to interact with those of plants in a way that led to a far more efficient and refined method by which pollen of one plant came to be conveyed very precisely to the female reproductive portion of another. An insect that is flying from place to place seeking mates or suitable places on which to lay its eggs is using energy, but is no longer able to replace this by feeding in its former way. If a plant were to produce a solution of sugar (i.e. nectar) in a position that was close to the structures that produce the pollen and the egg cells, and if the presence of this nutriment was advertised to insects by the evolution of structures (i.e. flowers) that were very noticeable, then adult insects would be likely to pass from flower to flower taking nectar for themselves; they would then also be inadvertently conveying on their bodies pollen from the flower of one plant to that of another; furthermore, by doing this they would be transferring it to the precise position the plant required it. This procedure, evolved as a result of natural selection, has worked excellently. It was far more refined and effective than the earlier methods that had depended on the vagaries of the wind. The results have benefited both the plants and the insects concerned.

As this strategy matured it also brought two other major advantages to plants. Firstly, the female gamete remained within the plant that had created it and so was able, after it had been fertilised, to receive nutritive material brought to it by that plant. This stock of nutriment would greatly help it when later, after it became detached as a ripe seed, it germinated as a seedling that had to fend for itself. Secondly, seed formation restored the opportunities for dispersal that had been lost at the time when spores of small uniform size (p.59) were replaced by the large megaspores that were retained in the parent plant. The mature plants could now not only supply nutriment to their seeds but could also, as a result of natural selection, evolve devices that equipped these seeds with effective measures of ensuring their dispersal. One instance is the nutritious fruit that surrounds the pips (i.e. seeds) of an apple; animals will welcome an opportunity to eat this fruit and, in doing so, will swallow the pips which, being indigestible, will pass unharmed through their digestive systems; these apple pips will therefore presently be voided, and very probably at a different place from where the apple tree was growing. Dispersal is thus achieved, and again in a way which benefits both the plants and animals concerned. A second example is the small parachute-like structures attached to the seeds of various plants, for example dandelions, which ensure that any wind will readily disperse them. A remarkably wide range of adaptations for seed dispersal have in fact been evolved.

The evolutionary emergence of these various interrelationships was a slow process. By about 80 million years ago it had reached a high degree of maturity. From that time onwards it has been the resultant 'Flowering Plants', presently existing in immense profusion and diversity, that have formed the principal component of the green mantle that has been covering the land surface of the Earth. Similarly it has been the more complex types of insects, with each part of their life histories specialised for quite different ways of living, that have been flourishing, likewise in abundance and diversity, along with the flowering plants that they have themselves been partly instrumental in creating.

During most of these last 80 million years the climate of the world has been warm and damp. Plant growth was therefore luxuriant, and the struggle to obtain sufficient light by becoming tall became once more a worldwide feature of the competitive scene. The predominant flowering plants were now in fact trees, and the principal vegetation was damp tropical forest. It was only comparatively recently, during the last ten million or so years, that much of the world's climate became colder and drier, and it was only still more recently that there have been acute 'Ice Ages' separated by milder inter-glacial periods. Hence it has been only in the equatorial regions of the Earth that these tropical forests, along with the abundant and diverse forms of life that they support, have been able to continue flourishing virtually unchecked. And it is these same forests, at present so full of future potential, that we humans are now so wantonly and effectively destroying.

Chapter 8

THE BRAINS OF MAMMALS

The last part of the previous chapter noted how flowering plants and insects came gradually to occupy most of the land surface of the world. This present chapter is concerned with the third major group of land-based organisms, namely vertebrates, which also took advantage of the environmental facilities that the increasing plant cover was providing.

Vertebrates, like plants and insects, required important modifications to enable them to pass from damp marshy lowlands to a drier environment. Fins that had already evolved into simple amphibian limbs needed further changes that enabled them to support at least a part of the weight of their bodies and so be used for walking. The skin needed to be thickened and adjusted to minimise loss of water by evaporation. Fertilization became internal and therefore independent of free water, and the eggs became large yolk-laden structures surrounded by a shell. This gave some protection, reduced loss of water by evaporation and provided enough nutriment to enable the embryo to delay hatching until it had become sufficiently developed to have a better chance of subsequently surviving. The evolution of characteristics of this kind, which made the animals more mobile, less subject to desiccation and largely independent of sources of free water, enabled them to become adapted to a whole range of terrestrial environments. Changes of this kind probably began during the Carboniferous, and perhaps among animals that were living in the intermediate zone between the swamps where coal was being formed and the drier land beyond. The more their evolution progressed along these lines the more effectively would they have been able to inhabit the new terrain.

A lizard basking on a wall today may give an indication of the kind of creature that would have been involved. It, like the lizard, would probably have been a fairly small animal with a long tail, and with a body covered with tough scale-bearing skin. It would have been alert and capable of moving quickly on limbs that barely raised its body off the ground. It would have been well able to find, seize and eat insects and small invertebrates. It, again like the lizard, would have mated and laid shell-covered eggs on land from which would have hatched juvenile animals similar in principle to itself. It would have had no system by which the temperature inside its body was kept constant and so, like the lizard, it would have been active when its body had been warmed by the Sun's radiation, and sluggish in the cold of night. To such creatures at that time the whole land surface of the Earth lay open to further potential evolutionary development, and this potential might have been expressed in numerous different ways.

The descendants of these reptilian ancestors split into two major groups. The first group gave rise to a number of subgroups that include lizards, snakes, crocodiles, tortoises and turtles. Representatives of these are still living at the present time. There are also a number of other reptiles of this major group that are now extinct; their former existence is known to us only by their fossil remains. These include dinosaurs, pterosaurs, ichthyosaurs and plesiosaurs, all of which included some very large and spectacular species. The dinosaurs dominated the surface of the land for about 150 million years, eventually becoming extinct, rather suddenly, 65 million years ago. The pterosaurs evolved means by which they could glide effectively, and hence could become airborne. And lastly the ichthyosaurs and plesiosaurs evolved, independently of one another, adaptations that enabled them to exchange their mode of life on land for one spent living in the sea; where they became very effective predators. There was also the famous fossil *Archaeopteryx*, of which a few specimens have been found in the lithographic limestone of Bavaria. It had rather simple feathers, and its structure suggests that it may have evolved from an early, small, fast-running type of dinosaur. Its feathers would have provided some thermal insulation, and so may have enabled its body to be maintained at a fairly warm temperature. Further elaboration of the simple feathers could have provided the means by which its descendants, or more probably those of its close relatives, eventually became able to fly, and thus to evolve into the very successful major group of vertebrates that we call birds.

The second main reptilian group, known as mammal-like reptiles, is extinct, and hence less widely known. In living reptiles, for instance lizards, the lower jaw is strengthened by a number of bones that are all roughly the same size. In the mammal-like reptiles one of these bones, the dentary, increased in size with the passage of time whereas the other bones became smaller and were relegated more to the back of the jaw. These mammal-like reptiles flourished for a long time and some of them became quite large by the standards of the time. They became extinct at the end of the Triassic period; this was about 200 million years ago, and thus at the time the evolution of the other major reptilian group, with its dinosaurs and so forth, was getting into its stride. At about this same time quite small creatures known as mammals appeared. In these the process of change that had been occurring in the mammal-like reptiles had reached completion, for in these mammals the lower jaw contained only a single bone, namely the dentary, and two of the other bones, now behind the jaw, have become members of a chain of minute bones that collectively help to transmit sound vibrations from the animal's ear drum to its inner ear. The mammals in question, and hence in due course all mammals, were presumably descended from a subgroup of mammal-like reptiles whose fossil remains have not yet been found.

It is this mammalian branch of the vertebrate stock that principally concerns us humans, perhaps not least because we ourselves are members of it. Throughout the Jurassic and Cretaceous periods, and therefore until about 65 million years ago, most mammals were quite small creatures. Decomposition was likely to result in their skeletons becoming disarticulated soon after death, with the result that fossil evidence about mammals that lived between 200 and 65 million years ago is largely limited to scattered bones and teeth. However, fairly recently, the obvious importance of the mammals living at that time has stimulated special studies of appropriate fossil-bearing beds in many different countries, and much new and significant information is now coming to light.

The other crucial source of information is of course those mammals that are still living at the present time. They enable us to try to interpret dry fossil remains, and thus mentally to clothe the creatures that they represent with blood, muscle and brain, and also with something of the zest for life that they would have experienced in their time. In almost all living mammals the posterior teeth are equipped with cusps, instead of being merely pointed structures, as in most reptiles. This enables them not only to hold food in their mouths, but to chew it, and thus to make use of tough types of food. Fossils show that cusps were present on the posterior teeth of even the oldest known mammals, which suggests that they also could deal in this way with difficult types of food such as seeds and insects, both of which would have been becoming increasingly plentiful. In the course of their evolution different groups of early mammals evolved different types of cusp pattern, these probably being adaptations to the particular types of food that its members had become accustomed to eating.

We know that the bodies of mammals living at the present time are covered with hair, and this was presumably the case also in the past. This, like the feathers of birds, provides some thermal insulation and so enables their bodies to be maintained at a constant warm temperature, which in turn gives them the advantage of being able to be active by night as well as by day. However it also means that in the case of these small mammals, as also for instance in shrews living today, body surface is necessarily large as compared with body weight, so a very high intake of food is required in order to compensate for loss of energy from this internal heating system.

In mammals fertilization is internal, and sperm can therefore reach egg cells without the presence of an external source of water. This is a very important adjustment or adaptation to the problem they shared with reptiles, of living on dry land. However mammals differ importantly from reptiles in that (with the exception of monotremes) female mammals do not lay their eggs; instead they are retained in her uterus or womb, and the resulting embryos are nourished by materials that diffuse from her blood stream into that of the embryos. This is a different, but effective, method of providing foetuses with the nourishment and protection they need. After the young are born the mother continues to provide them with nutriment, but now in the form of milk secreted by her mammary glands. This suckling of the young by the mother means that she spends part of her time keeping company with them, playing with them, protecting them, and thus also teaching them and mutually interacting with them.

An even more important feature of mammals concerns their brains. There is, as in all vertebrates, the basic portion of the brain which is responsible for generating such sensations as hunger, fear, pain and sex desire, and such instincts as, for instance, that which drives newly-hatched turtles immediately to run downwards to the sea. However in addition to this basic portion of the brain, and rather distinct and separate from it, mammals have evolved a pair of large dome-shaped structures known as cerebral hemispheres. Their surface region is tightly packed with nerve cells (neurons), and is collectively known as the cerebral cortex. In many mammals deep groves (fissures) in the cerebral hemispheres greatly increase the effective surface area, and thus the total area occupied by the cerebral cortex. The number of neurons in this cortex is impressive; in an adult human each square millimetre of its surface contains about 100,000 of them, and in the whole cortex there are about 10,000,000,000. Furthermore each neuron may, through its delicate peripheral branches (dendrites) have up to 10,000 functional connections with other neurons, many of these being close by, but some may reach quite distant parts of the brain. One can regard the mammalian cerebral hemispheres, with their cortex, as a new region of the brain that is primarily concerned with sorting out, categorising and interrelating the various aspects of the information it receives about conditions or events in its environment. This can be used, in collaboration with the basic portion of the brain, for determining what action the animal or person in question should take, and for putting this into effect. This type of cortex infrastructure, with its scope for forming innumerable networks of nerve activity that can mutually influence one another, has become well adapted for serving this kind of role.

The positions where information is passed from one neuron to another are known as synapses, and these transmission points are of critical importance. This applies not only to nerve networks, such as those in the cerebral cortex of mammals, but also to those of relatively simple invertebrates such as the mollusc *Aplysia* in which the entire nervous system consists merely of a few thousand neurons. In all such cases a point of special interest is that when a synapse frequently transmits information from one neuron to another, the structure of the synapse becomes modified in ways that enable it thereafter to transmit more readily. The change in the condition of the synapse is not merely temporary, but persists. The occurrence of this change, along with the circumstances that led up to it, are transmitted along nerve networks to reach and be registered by wide regions of the nervous system as a whole. In short, the experience will be memorised, and this memory can be used by other regions of the brain to help the animal in the overall business of its subsequent living. Thus an animal, whether it be a simple one like *Aplysia* or a mammal with highly refined cortical networks, can in some degree remember the experiences that it has had earlier in its life, and can modify its subsequent behaviour in that light.

It follows that in mammals memories are included in the stock of information that is available to this new versatile cortical region of the brain of mammals. An individual can continually accumulate experiences and learn from them, and on this basis add further to them, throughout its life. At all times its brain, or at all events its cerebral cortex, is by small successive separate steps gradually reorganizing itself in response to these experiences. Reorganization or remolding seem appropriate terms, for throughout this process a brain's various parts and activities remain closely integrated. We, who are mammals with particularly well-developed cerebral cortices, know the results of this as it applies to ourselves. Each of us can remember things and ponder on them; our capabilities are improved by practice; and our personalities do change in relation to our experiences as the years go by. Yet we do at all times remain a single integrated person, even though this person may at times feel somewhat ill at ease and divided within himself. Thus mammalian brains, and not least human ones, are at all times active, dynamic and potentially creative organs.

A mammal is able to relate to its environment in two ways. Firstly, the general structure of an individual's nervous system is predetermined by the pattern of genes with which that individual is endowed, and by the results of this in terms of the embryological brain development that follows from it. The broad outline of this pattern has been fashioned by the pressure of natural selection acting on countless generations of its ancestors. This is inherited, and the individual's initial instinctive responses will be an outcome of it. Secondly, the way this nervous system functions will be influenced by the experiences an individual has in the course of its own lifetime. Knowledge of such experience is not passed on from generation to generation; a son has no direct awareness of the experience accumulated by his father; these memories are, in other words, not inherited. In this respect each individual youngster has to start afresh, to experience and to learn from its own experiences, and also to learn to make appropriate use of what it has learnt. It is the first of these two features, namely the hereditary one, that predominates in simple animals like *Aplysia* whereas in mammals, with their complicated cerebral cortical networks, the second type, namely learning from experience, has also become exceedingly important.

It is also important that the cerebral cortex of a mammalian brain continues to grow until the animal has nearly reached maturity. Throughout a large part of the juvenile period of a mammal's life large numbers of new neurons are being produced, and this flood of new nerve-cells comes to be directed towards those areas of the cortex where synapses and memory generation are particularly active, and thus to those regions where further development is proving to be particularly important to the life of a young individual. These late-formed neurons therefore reinforce areas of the cortex where activity has already been accentuated in response to the environmental needs and experiences of the juveniles in question, The story is further complicated because, as already noted, a mammalian mother suckles and protects her young, and this provides her with an opportunity to try to teach her offspring some of the lessons that she herself has learnt from her own experience. In the contemporary human situation this need to try to help the young to learn from the experiences of their predecessors has been extended to cover the whole field to which the word education is now applied. One might add that the ability of the cortex to assess and compare experiences gives to a maturing mammal, whether human or otherwise, an opportunity presently to make, in the light of what he has learnt, his own self-determined choice about how he will conduct his life. He does not need merely to follow habits established by his forebears.

There are in a mammalian cerebral cortex a number of specialised regions each with their own particular function. At the back of the cerebral hemispheres there is, for example, a large area of cortex that is principally concerned with vision. Similarly at their front end there is a large area that is apparently concerned with interrelating the various types of information that the brain is continually receiving, thus presumably creating for the individual something in the nature of thought concerning them. Comparable specialization has been noted as giving rise to organelles in eukaryote cells (p.37-39), and to organs in multicellular animals (p.47).

Let us try to imagine what all this amounts to, using vision as an example. You and I, who are ourselves a mammal, are capable of 'seeing' a particular object, for instance a mouse. A mammal of quite a different kind, such as a cat, may also see this mouse. In each case photons of light radiated from the Sun impinge on the external surface of the mouse, and those that are not absorbed are reflected back into surrounding space. A small proportion of these latter reach the eye of the observer and, after being focused by its lens, impinge upon the retina of the eye. This 'information' is transformed by cells in the retina into the form of electrical impulses (i.e. nerve impulses), and these pass along the optic nerves to reach neurons in a particular region of the visual cortex. The medium is different, but the pattern that was initiated by the impact of light on the surface of the mouse is still retained. The 'information' that thus reaches a particular set of neurons in the visual cortex is then apparently relayed to other positions in this cortex; here local neurons analyse the pattern and deduce from it the relative dimensions (i.e. length, breadth, height etc.) of the original object, namely the mouse. In addition other neurons, equipped with much longer extensions, transmit the resulting information further afield to other regions of the observer's cortex, and in particular to the frontal area. The cortex of the observer will thus be made aware that something has been seen, and will in some way transform this information into a representation of the object. The nature of the representation will depend partly on the available information, but partly also on the type of cortex that creates the representation, and thus on the animal's earlier evolution, its concern about the object in question, and so forth. All this will differ crucially between a person and a cat. Each of them will view, or see, the same mouse differently.

In short, then, the only information available to the cortex is a pattern of nerve impulses. Yet what we see, or the cat sees, is something different and more useful. It is a mental image, or representation, of the type that we experience when we talk about having seen a mouse. The image seen by the man and the cat will both be based on the same basic raw material. Nevertheless the images will differ because the respective cerebral cortices, and also the attitude of these cortices to mice, will differ. We know from experience the kind of representation created by our own brain; on the other hand we have no means of experiencing, and thus of 'knowing', the one created by the cat's brain. The two images or representations doubtless differ widely, and presumably both will differ even more fundamentally from the actual mouse that exists 'out there', and does so irrespective of whether it is being looked at and by what. Both these observers form within themselves representations of the mouse that are appropriate to their own requirements. Seeing is so easy, and apparently so commonplace, that we are apt to forget about the unknowable difference that exists between a thing that is looked at and the impression of it that is formed within our minds.

The long period of relative tranquility during which dinosaurs had ruled the surface of the Earth ended at the close of the Cretaceous period 65 million years ago. All the species of dinosaurs living at that time then rather suddenly became extinct, as also did a very large number of species of other quite different kinds of life. The reason is uncertain. There is some evidence that a huge meteorite struck the Earth at this time. However it was also a period of intense volcanic activity, and the resultant gas and debris may have polluted the whole of the Earth's atmosphere and consequently grossly disturbed its climate for a long period.

In any case, the demise of the dinosaurs provided new opportunities for the mammals. One can visualise them as being, initially, small hairy warm-blooded creatures that had unspecialized hands and feet, that gave birth to their young and that had gradually evolved an effective form of cerebral cortex as a result of long evolutionary experience. These creatures would have been living in a rich and varied environment, with a wealth of flowering trees, shrubs and smaller plants, and of innumerable insects that these plants supported. They themselves would have been rather inconspicuous creatures that lived mostly in dense herbage on the ground and fed on insects, seeds and similar materials that they located largely by their sense of smell. The shrews that are living in the fields and waysides at the present time have retained many of these characteristics.

One can also assume that soon after the dinosaurs had disappeared some of these mammals moved into numerous different types of habitat that had consequently been left vacant. By doing this they would have been becoming divided into separate isolated groups each associated with a particular kind of environment and way of life, and each therefore soon developing into a separate species with its type of gene pool (p.39). Thereafter each such group would have pursued its own separate independent evolutionary course. The usual consequences of natural selection would in each case have had their different effects on its size, its hairy coat, its heat regulation machinery, the structure of its limbs, the details of its reproduction and mother-young relationships, and so forth. One can also assume that pruning by natural selection would have ensured that the various adaptations would have come to interact with one another in ways that would have given rise to well-integrated individuals that, both in their detail and their totality, were well able to function effectively in the type of habitat to which their species had become committed. Their cerebral cortices would of course have been similarly subject to the effects of natural selection. If, for instance, sight becomes important to the new way of life, then the visual area of the cortex will become large; if not, then it remains small. Evolution of brain and way of life thus also travelled together. The resulting new mammalian diversity differed from the earlier reptilian one in bringing to the land surfaces of the planet animals that were far more intelligent, each in different ways, depending on their particular circumstances and modes of life.

The results of this mammalian diversification have been remarkable. Today (leaving aside the monotremes and marsupials) there are in the first place the shrews and other insectivores that remain among the ground vegetation, still often eat insects, and have changed comparatively little. Bats have also changed little, apart from having evolved wings and a capacity for echolocation. There are primates; they moved up into the branches of trees where they feed mainly on fruit and insects. There are rodents, such as rats and mice, which feed on plants. There are small carnivores, such as stoats and weasels, which prey on the rodents. There are also two major groups of large herbivorous animals, one represented by horses, and the other by cattle and deer; their evolution has been associated with grazing on the extensive grasslands that developed when various regions of the world became drier and cooler a few million years ago. These two groups have independently evolved long slender limbs and the habit of walking on the tips of their toes, and are therefore well adapted to evading predators by running at high speed. There are also the large carnivores, such as lions, leopards and wolves, which prey on these large herbivores and have evolved in relation to them. And there are some mammals, such as seals and whales, whose ancestors, after living and evolving on land for millions of years, moved back into the sea and became adapted to that environment. Thus it would seem that the descendants of the small inconspicuous mammals of the Jurassic and Cretaceous, equipped with a cerebral cortex and corresponding intelligence potential, have made good use of the new opportunities for biological evolution that unexpectedly emerged after the demise of the dinosaurs 65 million years ago.

Chapter 9

THE MAKING OF HUMANKIND

The present chapter is mainly concerned with the evolutionary development of one particular subgroup of mammals, namely the primates, and with one species within that subgroup, namely *Homo sapiens*. The story begins shortly after dinosaurs became extinct about 65 million years ago, when many of the non-specialised mammals that had been living among the herbage on the ground started moving into habitats that had now been left vacant. The group that eventually became primates moved upwards into the branches of the shrubs and trees above them. Here they would have found sufficient insects and seeds on which to feed and would have evaded predators that were left behind them on the ground.

The potentialities afforded by this way of life were great, for the climate of much of the world was then warm and damp, and much of its land-surface was covered with forest. In this new arboreal habitat vision was more important than smell, for these small creatures needed to be able to run dexterously along branches, and to jump from branch to branch. Under the influence of natural selection their eyes and the visual portion of their cerebral cortex evolved in ways that enabled them to perceive the precise positions and shapes of the branches that they traversed. Similarly their hands and feet evolved into structures that were highly sensitive and supple, and could therefore conform readily to the shapes of these same branches, and so grasp them firmly. Lastly they evolved a form of brain organization, and in particular of cerebral cortex, that enabled them to translate the information they received through sight into movements that were sufficiently precise to enable their hands and feet to make appropriate contact with the branches where they ran or jumped. The activities of hand, eye and brain thus became very well coordinated. We present-day humans are descended from early tree-living primates and have inherited some of these characteristic; we find them invaluable on innumerable occasions in the course of our daily lives.

Living among the branches of trees also had other consequences. A mother who is jumping from one branch to another cannot possibly look after numerous young, and in most primates the number of offspring in a litter has been reduced to one. Also each youngster has much to learn. Its genes have been coded to produce an individual with refined powers of eye and brain and hand, but it will have to learn to make use of this equipment by practice during childhood. Learning is a slow business, and the length of time for which a youngster is actively assisted and instructed both by its mother and by other members of its group has tended to increase in the course of primate evolution. In our human species it has, of course, become particularly long.

Most species of primates have apparently found it beneficial to live in mutually cooperative groups. The search for fruit is one activity in which members of a group can help one another. In a tropical forest an occasional tree will come into fruit at a particular time and will then provide an abundant, but temporary, supply of food. Such sources are sporadic and widely scattered. When one member of a group discovers such a tree it can call loudly to proclaim the fact and thus attract other members to the spot. There is plenty of food there, so the communication of the discovery, and the sharing of the bounty, is to the mutual benefit of all concerned. In such ways bonds are formed between the members of a group.

Another result of living in trees is that in many primates most of the weight of the body is borne by the hind limbs; the forelimbs are then used mainly to assist balance by clinging to adjacent small branches or twigs. The animal's posture has thus become semi-upright. This means that its very capable hands can then be employed not only for locomotion but also on occasion for such tasks as probing or investigating insects or fruits, and then conveying those regarded as suitable to its mouth. The latter therefore no longer had to be able to seize and hold food, and the jaws could become shorter. This in turn facilitated the evolution of eyes that were directed forwards, and in fact vision in these animals became stereoscopic.

These various developments naturally overlapped and mutually affected one another. Primate evolution has therefore taken the form of an evolving network of interacting characteristics. The overall result has been the coming into existence of dexterous creatures that usually live in more or less cooperative groups, have excellent vision, and are curious about objects and well able to manipulate these with their hands. Such characteristics are demonstrated in a relatively simple form by lemurs and tarsiers, and to a greater degree by monkeys, as also likewise by ourselves.

Fossils provide an important source of information. Of particular interest are the numerous fossil remains of small apes (dryopithecines) that lived in the warm forests that still covered so much of the world's land surface about 20 million years ago. Clearly the primate mode of living was very successful at that time. Later the climate changed somewhat, becoming colder and drier, and large areas of forest were replaced by grassland. Fossils of this type then become less numerous, and have not been found in rocks that were formed more recently than about 8 million years ago. These fossils are also interesting because their structure shows that animals of this type may have been directly ancestral to the two types of apes at present living in Africa, namely chimpanzees and gorillas. These differ from the earlier apes mainly in being larger and heavier animals which are inclined to use their arms for swinging from one branch to another. It is also probable that the so-called hominid branch of primate evolution, which by forsaking the trees and returning to the ground led eventually to humankind, was also descended from these earlier apes. In this context it is relevant that the genes of chimpanzees, and also to a lesser extent those of gorillas, are exceedingly similar to those of human beings.

The earliest known fossil remains of the ground-based hominid branch of primate evolution are relatively recent, dating merely from between four and three million years ago. They have been found near Hadar in northern Ethiopia, and at Laetoli in Tanzania, both of which are within the Great Rift Valley system of Africa. The Ethiopian material includes a considerable part of the skeleton of a small adult known unconventionally as 'Lucy'. These fossils indicate on the one hand that the brains of these primates (when differences in body-size are taken into account) were very little, if any, larger than those of the early fossil apes or of chimpanzees; also their teeth were very similar to those of apes except that their canine teeth were shorter. On the other hand their posture was almost fully upright and their feet, being well adapted for walking bipedally on the ground, were less well suited for climbing trees. At Laetoli there are similar jaws and teeth, and in addition footprints that provide dramatic confirmation of a bipedal gait. Thus it would appear that the crucial difference between the two primate lines that led respectively to modern apes and to humans was that the members of the latter had exchanged their former life in trees for one now spent on the ground. Furthermore this change had probably taken place considerably more than 3 million years ago, since by that time the bodies of these hominids had apparently become fairly well adapted to the complex locomotory requirements involved in bipedal walking.

It would be interesting to know why these apes left the trees. The reduction in the area covered by forest may have been an important factor. By descending to the ground and moving onto open grassland they would have been entering an environment broadly similar to that of the East African plains at the present time. It would have offered a wider range of food than the trees, but, on the other hand, potential predators would have been more numerous, and we do not know how they dealt with this problem.

There is as yet little fossil evidence regarding hominid evolution during the period between three and two million years ago. However for that between two and one million years there are a number of very important fossil sites, again mainly in the Rift Valley, and more particularly from Olduvai Gorge in Tanzania and Kobi Fora in northern Kenya. Here volcanic activity associated with the Rift Valley (p.25) led periodically to the deposition of layers of volcanic ash and lava. After this had cooled it provided a covering that helped to preserve sediments and associated fossils, including the Laetoli footprints, from erosion. Certain minerals in these volcanic rocks provide a reliable means of calculating the approximate time the sedimentary beds adjacent to them, along with their fossil content, were formed.

It seems that by two million years ago, and therefore at the beginning of this period, these ground-dwelling primates i.e. hominids had diverged into two separate evolutionary lines, known as australopithecines and hominoids. The brains of the former remained about the same size as those of chimpanzees. There is no evidence that they made tools. They became extinct about one million years ago.

The other sector, namely the hominoids, are more interesting in the present context. They may have comprised a single rapidly-evolving evolutionary line. Their most striking feature is the rapidity with which the size of their brains increased, from an average volume of about 600 c.c. (only slightly larger than that of a chimpanzee) about two million years ago to about 1000 c.c. by one million years ago, and always with a very wide range of variation between one individual and another. Fossils which belong to the less developed end of the series are referred to as *Homo habilis*, and those at its more developed end as *Homo erectus*. A new departure was that they chipped stones into shapes that could be used as tools, so enabling them to do things that their unaided hands could not possibly have done. At the beginning of this period these tools were crude, but by its end they had become far more refined. They lived in small groups or communities, and some of them established for themselves living sites or home bases which they occupied for at least some considerable period during a year.

About one million years ago some groups of *Homo erectus* were beginning to move away from their African homelands. Discoveries of their skeletons, their tools and in some cases their living sites show that they gradually extended their range northwards to the Mediterranean, and from there further north into Spain and Hungary; they also moved eastwards into Asia where they reached Java and Beijing. They engaged in complex cooperative activities for obtaining the meat of large animals, for instance by driving elephants over cliffs in the south of Spain. They made clothes from the skins of animals they had killed; without these they could not have moved northwards into the much colder climates. They also discovered how to make use of fire.

By about 100,000 years ago these hominoids, whether in Africa or elsewhere, had a brain that was on average as large as that of ourselves (i.e. varying between 1200 and 1600 c.c.). However they still retained the rather thick bones and prominent brow-ridges of their earlier *Homo erectus* ancestors, and are often referred to as belonging to the Neanderthal form of the species *Homo sapiens*. Somewhat later fossils demonstrate the presence of populations in which the bones and brow-ridges were relatively slender. Thus at this stage, about 40,000 years ago, there had come into existence a type of living organism that seems in its physical characters, and therefore probably also in its inherent mental abilities, to have been virtually indistinguishable from ourselves. The mental sophistication of early humans is demonstrated by the sculptures and paintings of animals that they were already making in caves some 20,000 years ago, and of their stone tools, which had become refined and delicate, with each type made to a definite pattern and for a definite purpose.

There is now evidence that suggests that this new and more slender type of *Homo sapiens* was not descended from the Neanderthals who had been in Europe and Asia, but was a product of a further evolutionary development in the African heartland which had then, like its predecessor, spread from there into Europe and Asia. This time they traveled further, reaching the American continent by way of the land bridge that then existed in the region of the present Bering Straits. Also, much further south, they reached Australasia. Thus it appears to have been these more slender types that eventually populated the world with its present type of human beings. This would explain the very small genetic differences that separate the present races of humankind. Essentially we all belong to one single human family.

There are of course numerous aspects of biological evolution in general, and of human evolution in particular, that leave no fossil evidence; we are then wholly dependent on information provided by the nature of the humans who are living at the present time. Speech is an example. We know from our own experience that we can speak. We also know that an ability to speak is common to all humankind, though the language used differs from one community to another. Naom Chomsky has stressed that this potential ability to speak is inherited, a small area of our cerebral cortex having already become organised before our birth in a way that makes speech possible. Learning to bring this possibility to fruition, and in the language used by the community into which a baby happens to have been born, is of course another matter. It depends on the development of appropriate mutual interactions between the baby and its mother, and also between it and other members of the community in question. In other words, it has to learn to speak the language of the community in which it lives.

As already noted, most primates live in loosely organised groups or communities whose members can communicate with one another vocally, as for example by shouting with elation on the discovery of a fruit-bearing tree. The information exchanged is however largely concerned with emotional states. Human speech, by contrast, is more complex, and the information conveyed is far more precise. Particular sounds, referred to as words, are used as symbols to represent particular concepts. This probably requires as a precondition not merely an awareness of the existence of categories of things (such, for instance, as 'dogs' or 'things that are red'), but in addition an appreciation of the likenesses between all members of the same category, as for instance all dogs, or all experiences of the colour red. Once such categories are represented by concepts created in the cerebral cortex then it becomes possible in principle for these same concepts to be symbolised by particular words. such as 'dog' and 'red'.

However the production of sounds that were sufficiently precise would doubtless have posed a problem. The motor region of the primate cerebral cortex had long been able to instruct the limb muscles in ways that gave rise to precise and purposeful movements of the hands and fingers. The utterance of well-articulated words, and also the formation of sentences, would now have required the gradual evolution of a comparable system that activated the muscles of the vocal cords, and of the tongue and lips, with extreme precision. This process would doubtless have been gradual, slow and difficult. However each success would have been helpful in itself, permitting more effective communication, and so encouraging and making possible further progress in the same direction. In any case we know from our own experience that this evolution was in fact accomplished, for today we inherit an ability to learn to speak.

The primary function of speech was presumably to provide a means of precise communication, but it has also had some very important secondary consequences. It has enabled information and associated ideas to be exchanged so that they flow between the different members of a community, who can thus come to share a common climate of conceptual thought. It has meant that concepts that would otherwise have existed only vaguely in a person's mind become clarified, categorised and organised as an inseparable part of the process of learning to speak. Also the words that a child is taught will be relevant to the society in which he is brought up, and the learning of them will help to attune him to its concepts and its ways. Words also, very importantly, provide symbols which a person can use to think about things in their absence, and he can therefore allow these symbols, and the ideas that they symbolise, to interact freely with one another in his mind. Words therefore add a new dimension to opportunities for thought, greatly adding to the scope for creating new patterns of cerebral activity, and for integrating old ones, and hence in general for developing new insights. Some of these tentative new concepts may be relevant to the time, place and circumstances of the community in which he lives, and their potential value in this context can be tested. There will also be others that are more imaginative in the sense that they have no obvious relevance to immediate problems. Both these types of intuition may, when followed up, lead to whole new fields of human experience that would never have been conceived without the use of words as symbols. Throughout such evolutionary processes brain and speech and cooperative group activity will have been mutually interacting, with each positively encouraging the further development of the others. It was presumably activities of this kind, being acted upon by natural selection, that led during the last two million or so years to the exceedingly rapid increase in hominoid brain size that has been noted.

At about the same time as speech was being created, and probably both aiding its development and being aided by it, there was a tendency for men and women who had previously been wholly nomadic to begin to establish living sites or home bases that they would occupy for at least some considerable portion of a year. Such sites were widely scattered. They might be beside some river or lake in Africa, or at the entrance to some cave in a Europe that was at that time afflicted with the last major phase of the Ice Age. Much information has resulted from careful studies of bones, tools and other remains that had gradually accumulated at these sites.

One can visualise these home bases as busy places where some of the occupants would have been tending fires, preparing stone tools or making clothes from the skins of animals. Here also the adults would have discussed the events of the day and planned the work for the morrow; for the men this might have been about the coordination of their hunting procedure in the surrounding area, and for the women about the gathering of fruits, seeds, roots and perhaps firewood. In both cases the products would have been carried back to the living site, and there shared. Here also there would have been the cooking of this food, and the eating of it. It was a place where the children would have played, and learnt to speak, and become attuned to the ways of the group. The base was therefore a small localised area in which many interrelated activities were taking place more or less simultaneously, and in which men and women, using their intelligence, their energy and their capacity for cooperation, were the organizing agents.

One can also view a living site as a minute sector of the world in which the normal ecological relationships that functioned elsewhere in the vicinity were no longer operative. Instead there was here a miscellaneous collection of equipment, such as hearths and toolmaking facilities, and of course the humans themselves who were using the site merely to promote their own immediate interests. Outside this small much modified location there was, by contrast, the surrounding natural environment. This was as yet still vast, extending in all directions around these minute home bases. To the people who were living in them the environment would have been regarded as the place into which they would go out to gather and bring back the food and other materials that their small community required for the business of its living. Present-day hunter-gatherer communities have, as a result of their day-to-day activities, become remarkably well acquainted with the details of those aspects of their environment that concern them. Likewise they are very conscious of the vital importance of conserving that environment while at the same time drawing from it the materials they require. They appreciate that their own lives, and the futures of their children, depend on their succeeding in doing this satisfactorily.

This new kind of system, a human social system, comprising the home base and the persons who had their home there, differed from other systems previously noted, such as atoms, molecules, living cells and multicellular organisms, in a number of important ways. In the first place, unlike atoms and molecules but like living organisms, it was of course an open system; the energy needed for its maintenance was drawn from outside itself, as radiation from the Sun, as food, and as fuel. Nevertheless its outer boundary was indefinite and immaterial; there was nothing that corresponds to the membrane that surrounds a living cell, or the skin that delimits the body of an animal, nor was there any need for such. Also, partly for this reason, it was less integrated; the humans would not have organised everything at the home location, but only those aspects of it that concerned them. Some plants would have continued to grow there, flies interested in the food there would have flown in and temporally joined the scene, and perhaps the occasional dog, as yet undomesticated, would have looked round for discarded scraps. Comparable intrusions do not usually occur within the confines of an atom, or of a cell, or of a human or non-human animal. These home-based social systems were thus less firmly structured, and less integrated.

Secondly, by means of speech, the individuals in a society could achieve an inter-penetration of their thoughts, and they could therefore work collectively towards a common aim that would of course include the maintenance and the wellbeing of themselves and their society. The contact between brain and brain needed for social living was therefore subtle; speech could provide it, and it involved the individual in no physical or biological restrictions. The intelligent human units could thus organise and be a part of the organization, and yet move freely through this organised environment, which therefore was a medium in which they could also live private lives. This was quite a new kind of relationship; in a multicellular organism, for instance, the individual cells, or parts, have no such freedom and the whole is far more closely integrated. One consequence was that the individual human had scope for promoting his own self-interest as well as a need to participate in the maintenance of a society that had become essential to his existence. This dualism has had very important implications, and it lies at the heart of some of our difficulties at the present time. There is always a potential conflict between immediate personal interests and wider or longer term considerations of a more public kind.

A third difference is that a society is potentially immortal, whereas its constituent human beings are mortal. This mortality aspect has a de-stabilising affect. It means that at all times young people are taking the place of those who die; moreover the society in which these young people grow up forms an important part of their environment. It follows that if the nature of a society changes, for instance as a result of the introduction of some important new technique, the pattern of the cortical network that develops in these young persons' brains will become modified in response to this innovation. This in turn will affect the attitudes of the young people who grow up under its influence. This may give rise to further changes in society, which in turn may affect the next generation of youngsters, and so forth. This type of continuing mutual interaction between social environment and its mental counterpart can lead to cultural evolution taking place very rapidly. However it is important that these cultural changes do not affect the genes of the persons concerned, and are therefore not inherited. It is only after our birth that our social or cultural environment will affect the type of human beings that we will eventually become. This separation between genes and environments moderates the rate of change of social evolution; nevertheless this usually proceeds vastly more rapidly than biological evolution.

This inevitable succession of generations presents a problem to contemporary hunter-gathering communities, as doubtless it did also during that very long earlier period when all humankind lived by hunter-gathering. Here the individuals find themselves members of a community that has its traditional routines; certain things are done regularly, and in certain ways. In the absence of writing, and hence of knowledge of their own past history, they know little or nothing of the origins of these traditional procedures. They do, however, know that their system does work, and they feel, not without reason, that any departure from it may bring disaster to their small and frail communities. Their spontaneous response has been to conceive some mythological version of the origins and history of themselves and their traditions, this probably involving some type of gods or spirits. This gives an explanation, or at any rate a rationalization, and thus provides them with a substitute for the history that is lacking. The adults will have learnt this version of the origins of their communal traditions while they were children, and they will in due course teach it to their own children. Belief in such myths is periodically fortified by appropriate ritualistic ceremonies. These mythological systems can provide answers to all possible questions, and in such a form that they cannot be falsified. They bring cohesion to their community, and help to keep it continuing in its same traditional ways, which they hope are safe. These societies therefore tend to be conservative.

The evidence suggests that the emergence of humankind required no special 'Act of Creation'. It was ordinary evolutionary processes, with their creative tendencies, that led to this result. A crucial factor was that the responses of our very distant ancestors to life among the branches of trees were not of a specialised kind, and therefore did not restrict them to this one particular mode of life. Their hands and feet were well suited for climbing trees, but were not restricted to that function by, for example, the development of claws; they could therefore be used, whether then or later, for other purposes, and have come to do so excellently in their respective ways. Likewise the detailed coordination between hand and eye and brain could be made to serve many different purposes, and not only in trees. The brain itself, with its tendency to promote manipulation, exploration, curiosity and enquiry, later evolved into a very effective general-purpose organ.

On the ground, the most critical period seems to have been that associated with the extraordinarily rapid increase in brain size between about two million and 100,000 years ago. A number of factors were probably mutually reinforcing, so giving rise to a positive feedback situation. These could have included the differential advantage as regards procreation that would often have been conferred on those individuals that showed themselves particularly well able to communicate, organise, make things, and in general to contribute to the proceedings of their group. This, in a setting of very small gene pools, could soon have led to a creation of gene assemblages that promoted a rapid increase in all-round brain ability. Other participants in this positive feedback could have been mutual interactions that were related to the evolving concept-based speech, the improvements in comprehension that this would have brought, along with the resulting imaginative insights and the testing of their practicability. This, when linked with their capacity for manipulating objects, could have resulted in new tools, new ways of doing things and new things to do. All these would have been parts of an interacting, evolving network that would have been becoming more complex and integrated as their brains became more effective. And time, thousands of years of it, was on their side.

It seems that the type of brain that was most useful, and so was evolved at this stage, was a very able general-purpose one. It needed to be able to deal effectively with the wide diversity of problems that hunter-gatherers were likely to have had to face during their full and complex lives. It could do this because its cortex could generate patterns of nerve activity relevant to the data it received. and because it could memorise, synthesise and thereby gradually create appropriate representations of human experiences. It is this potentiality that has enabled human beings, with their human brains, to participate successfully in exceedingly complex cooperative activities—such as making orchestral music or journeying to the Moon—which they were certainly not called upon to undertake when the type of brain in question was being most actively evolved. Thus the brains that we happen to have inherited are extremely versatile organs, with immense potential, much of which normally remains unused.

These human brain developments have, in their turn, opened new horizons. They cause each one of us, as we grow up, gradually to become conscious not only of the world that we experience around ourself, with its categories and relationships of things, but conscious also of ourself as one particular entity therein—i.e. the one we knows as the 'I' that is ourself—and conscious therefore of our own particular position within that world. Once this self-consciousness has been attained we can begin to appreciate the whole person-environment complex impersonally, as though viewed independently from outside ourself. The 'I' of others is then seen to be no less valid, and no less valuable, than that of ourself. Ethics can thus acquire a rational basis. Further, we knows that we can not only think about this complex, but can also act in relation to it, and thus affect it. We may find that, as regards some situation, several possible courses are available. We will then have to make a choice, or, in other words, use our own free-will. Our decision will necessarily be related to our overall appreciation of the situation, which in turn will be related to our earlier experiences, and to the insights that have resulted from our thoughts about them. It is our responsibility to use our capacity for careful thought, and so to choose as best we can, after giving due consideration to the ideas regarding relative values, ethics, appropriate attitudes and so forth that have become established in our mind. In other words, these evolutionary developments have enabled him to become an ethical animal, and a spiritual one, and possibly also a religious one. However, as we have seen in earlier chapters, a development of increasingly high levels of creativity, and a corresponding birth of qualities that are fundamentally new, has been a typical feature of evolution in general. In this case the fundamentally new characteristics have been, for example, self-consciousness, ethics and spirituality. There is nothing particularly divine about these developments. On the contrary, a gradual step by step transcendence of previous limitations seems to be an inherent feature both within the Universe in general and on planet Earth in particular.

The developments during the last two million or so years have affected the relationship between us human animals and the non-human animals around us. Human brains have become, relative to body-size, far the largest, and our cerebral cortex far the most complex, of any mammalian species; hence our capacity for thought has become the most effective. Also our ability to speak has enabled us humans to think more precisely, and our communities to act together more concisely, and so has brought to us whole new ranges of opportunities. Some non-human animals do think, communicate, manipulate objects, live in social groups, form home bases and engage in communal enterprises, but not one of them does all of these things nor does more than one or two of them effectively. In early humans all these types of activity were developed, and to a high degree; in addition they were coordinated in a way that gave rise to integrated systems that would have gradually matured at their home bases. Other animals had nothing equivalent to this effective interlocking network of abilities that could, in principle, be applied to almost any situation. Thus our ancestors gradually gained a fairly complete ascendancy over the larger kinds of animals, but not over smaller ones, such as mosquitos, ticks and fleas. During this period a number of species of large mammals did become extinct, and humans may perhaps have been responsible. Nevertheless for so long as these humans still continued to live by hunter-gathering they, like other animal species, were merely one of the components that collectively comprised the ongoing natural ecosystem that had gradually evolved in the course of some 4000 million years. Rather recently, a mere 10,000 years ago, this unusual species abandoned hunter-gathering, and so initiated a new and very different chapter in the history of this planet.

Full moon over a wheat field   Telegraph Colour Library, Benelux Press

Chapter 10

HUMANS MODIFY EARTH: EARTH MODIFIES HUMANS

The hunter-gatherers who became fully human some 40,000 years ago retained essentially this same mode of life for a very long period. An early indication of impending change can be seen about 12,000 years ago when some of their communities began to pay increasing attention to herds of docile herbivorous animals, such as goats and deer. They would keep in contact with these herds, perhaps to some extent protecting them from predators; they were thus treating them as a potential source of food that would remain available and in good condition until required. The animals, for their part, would have become accustomed to these humans being in their vicinity.

Changes of this kind may have been widespread at this time. There is evidence that they were taking place in the Pyrenees, and also in the upland area known as the 'Fertile Crescent' which stretches from what is now northern Palestine and Lebanon in the west to the border of Iran in the east. In this region of southwest Asia wild grasses with edible seeds i.e. wild cereals were growing (and still do grow) in abundance in suitable places. A people known as Natufians were living in the western portion of this crescent, and investigation of their living sites has shown that they used gazelles as semi-domesticated animals; also, importantly, they harvested these local cereals, which provided them with an additional source of food. They reaped them with sickle-shaped stone tools that they made for this purpose; the silica in the grass stems gave these sickles a characteristic sheen which is still visible today.

There is definite evidence that by about 10,000 years ago people in this region were not only herding animals and harvesting wild cereals, they were also sowing and cultivating the latter. One can guess that initially the regular gathering of wild cereal seed would have drawn their attention to the life history of these plants and that, using their intelligence and imagination, they would have conceived the idea of scattering some of this grain on damp ground and perhaps covering it with a little soil, so in effect sowing the seed. They would have found that this worked, for a concentrated crop would have resulted in due course. One can also surmise that, although this was a departure from their traditional routine, it would have seemed merely complementary to their normal activities, and would accordingly have been accepted as a beneficial innovation.

In fact, however, the consequences were to prove of crucial importance. Even though the crops themselves were only seasonal, the plants that generated them were present in the soil at all seasons, so weeds and grazing animals had to be continually kept away. People had to remain in the vicinity all the time, which meant that the home bases or settlements (p.80) had to become permanent. Groups of simple houses, comprising the beginnings of agricultural villages, were presently being created.

The great merit of this revolutionary technique was that it resulted in the production of more food, with greater certainty. Women in permanent dwellings could deal with a more rapid succession of babies than women who may have to move from one place to another. It became possible to rear greater numbers of children, and the size of these local agricultural populations increased accordingly. This in turn required bringing more land under cultivation to feed the larger community, which resulted in further population growth, and so forth. A positive feedback, or self-reinforcing situation, had been created. A stage was reached that any further suitable land that remained uncultivated was too far distant from the village to be worked efficiently. Some villagers would then have hived off to start a new settlement in that area. Further growth would have resulted in an increase in the number of village communities instead of a continuing increase in their size.

In addition there would have been an animal husbandry aspect to this agricultural revolution. There would, for example, have been times when more grain was produced than the people of a village needed, and the surplus could be fed to semi-domesticated animals. These animals would thus be more fully assimilated into this new type of people-made environment and so would, before long, become fully domesticated. One consequence was that thereafter they, like the plants, would require some human attention at all times of the year. Their manure would have enriched the soil, so increasing the yield of future crops. All this would have been a part of an overall stabilising self-reinforcing situation.

The scene would also have been changing in less obvious ways. Very large numbers of ripe seeds would have been produced at each harvest. The people, when they were selecting from these the seeds they would use for sowing, would have been likely to include any deviants that were particularly large, or

that seemed especially suitable in other ways. They would then have been selecting perhaps quite rare mutations for the purpose of breeding, and would
have been putting into effect not 'natural selection', which tends to promote
all-round efficiency, but would instead have been imposing an artificial
human-oriented selection that would, they hoped, prove beneficial to themselves. We know that something of this kind was happening, because the remains of seeds found at living sites became considerably larger and also differed in other ways from the original wild stock. Comparable changes are seen
in the bones of the domesticated animals; people would naturally have tended
to use those particular individuals for breeding that had been found to meet
their own human needs especially well.

This new agricultural mode of life naturally gave rise to new secondary activities. For example, the grain that was harvested had first to be brought in
from the fields, and then stored. Containers made of pottery could serve both
these requirements. Pottery was not altogether unknown to hunter-gatherer
communities, but now the means of producing it had to be explored, elaborated and utilised on a considerable scale. The first sorts of pottery apparently
seemed dull, soon people decorated it in ways that enabled them to express
their artistic tendencies.

This agricultural revolution was clearly very comprehensive, and it soon
began to change the overall condition of the world. In the preceding hunter-
gatherer societies the area in which the prevailing ecology was seriously disrupted by the presence of a living site was very small. Beyond this the natural
wilderness, with its ecological balance, remained virtually undisturbed. In an
agricultural setting the area of disturbance was much larger. There would, firstly, have been an inner core, corresponding roughly to a hunter-gatherer living site, but larger, for it would now comprise a number of permanent houses,
each perhaps occupied by a single family. In addition, and far more important,
there would now be a much larger peripheral zone beyond this village consisting of a new kind of creation, namely cultivated ground or fields. Here appropriate seeds would have been sown, and the growth of the young plants encouraged. Wild plants that competed with them would have been designated
weeds and as far as possible removed. Both the crop plants and the equivalent
animals doubtless soon became intrinsically unstable, for the gene assemblages
that enabled them to hold their ground in a natural ecological community
would have been altered by the selective breeding that was employed in the
new farming practice. As already noted, these agricultural developments were
self-reinforcing, so the area that was cultivated, and was therefore ecologically

unstable, steadily increased at the expense of the natural stable wilderness. It also followed that the remaining hunter-gatherers were steadily pushed outwards into less fertile and more difficult regions.

As a result of their adoption of this new way of life, the outlook of the people concerned would doubtless also have been gradually transformed. They would have had to organise a wider area and in a more complex manner, and to plan on a longer time-scale that recognized that the seeds they sowed today would not yield a harvest until a rather distant tomorrow. Also their concerns would have shifted away from the kind of natural history approach appropriate to the skills required for hunting and gathering, and have been replaced by a more biological one associated with the well-being of the plants and animals that they now tended. They would therefore have become concerned with such matters as the selection of seed to be used for sowing, with the placing and timing of that sowing, and with maintaining the resultant plants in a healthy condition. Similarly they would have been concerned with the feeding, the habits, the matings and the procreation of the animals they had domesticated. They would also have been engaging in quite new activities that had now become relevant, such as the making of pots and the baking of bread. They would probably also have differed from their hunter-gatherer predecessors in being conscious of having invested much time and effort in developing their settlement and its surrounding fields. They would therefore wish to maintain this property, and if necessary to defend it, and thus to continue to benefit from the arduous work that they had undertaken. Individual families would perhaps now have been regarded as owning the plots that they had developed, and so the tendency to share, which seems to be characteristic of hunter-gatherer communities, may have become less freely operative. Thus with the first sowing of seeds of cereal grasses they may also have been ultimately sowing the seeds of a more individualistic, property-conscious and acquisitive society.

The emergence of agriculture thus provides an early illustration of an important and widely relevant relationship, namely that changes in human personalities are closely related to changes in the environment that they experience. Each is continually affecting the other. Both therefore need to be viewed conjointly, not separately, for they are interacting parts of a single developing system.

The second important point is that subsequent to the emergence of agriculture the principal changes in human evolution have not been of a biological or genetic type, but of a cultural or social kind. These cultural developments are not inherited, and therefore each new generation has to adapt to them afresh. This type of evolution is vastly more rapid than its earlier biological counterpart, and has indeed now reached such a speed that we humans, who initiated it, are finding it difficult to control its course. The next four chapters are intended to indicate its approximate course so far.

Pyramids at Giza, Egypt   Grant V. Faint, The Image Bank

Chapter 11

EARLY CIVILIZATION

The agricultural revolution referred to in the last chapter began about 10,000 years ago, and it is probable that during the next four thousand years most of the suitable land in the Fertile Crescent and adjoining regions was brought under cultivation. The resultant lack of further land may have provided the incentive that led the Sumerians, about 6000 years ago, to undertake the clearance and irrigation of a portion of the adjacent reedy, marshy ground that lay between the Tigris and Euphrates rivers. The completion of this arduous task rewarded the Sumerian communities with a soil that, in addition to being very fertile, had this fertility periodically renewed by fresh silt deposited in time of flood by the two rivers. It was able to produce, and to continue producing, crops very much larger than any normally experienced. Thus an ordinary agricultural community proved able to create from unpromising land an area that became extraordinarily productive.

One result of this extreme fertility was that the number of people in a local population could go on increasing until it reached a remarkable density. This had numerous consequences. For instance, cultivation of the soil required the work of only a small portion of this much larger community. The people thus made unemployed were available for other types of work which were becoming available as a result of the changing circumstances. Some became engaged in transporting surplus grain to neighbouring communities, and in trading this for copper, silver and other metal ores, and for timber, none of which were obtainable in any quantity on the alluvial plains on which they lived. These imported materials were then be used for productive work at home. In this way, in addition to the people who tilled the fields and maintained the irrigation system, carpenters, builders, and smiths who worked with copper, silver and gold came into being. There were also the merchants who left home periodically to trade in neighbouring lands, and the soldiers who might accompany their convoys, or who might protect the accumulating wealth at home from others who might covet it. All this was an initial result of the good use they had made of the exceptional agricultural potential of their land.

The administration of the various very small townships that grew up in Sumer was undertaken by their respective temple authorities. They used the outbuildings of the temples as centres for the redistribution of goods and for the general organization of the economy. They collected taxes in kind, and they were responsible for using the grain and other materials that they thus received to provide the merchants with goods that they could export, and similarly for using the imported metals and timber to provide the craftsmen with the raw materials that they needed for their work. It is significant that the temples, which in the early stages were quite small, were within a few centuries transformed into relatively huge structures, known as ziggurats, which towered above the other buildings in the townships.

The people who consequently spent their lives doing these different types of work would have been experiencing different types of small-scale environments. The nerve networks created in their cerebral cortex would accordingly have differed, so that the people who spent their working lives engaged in each such occupation would have come, on average, to develop corresponding distinctive types of personality, outlook, interests and economic status. There would thus have been an early form of class differentiation. Further, since in most cases the children would have necessarily followed in their fathers' footsteps, these non-inherited environmental differences would have been perpetuated, so giving rise to enduring social classes. These complexities would have brought into existence forms of social organization and relationship that were very different from those experienced by hunter-gatherer or simple agricultural communities, where all persons did much the same kind of things and experienced much the same environments. These communities had much the same common interests and outlook.

The new situation raised new problems, and seems at first to have encouraged people to discover new ways of doing things. The various techniques being developed interacted with, and stimulated, one another. Plenty of copper could be imported; this facilitated the discovery of the method of casting known as *cire perdu*, and this enabled copper and bronze to be cast into precise predetermined forms, which in turn made possible the production of more effective tools for carpentry, such as metal saws. These could in their turn be used to fashion wood into circular shapes that could be put to use as wheels for carts, and then these carts, drawn by animals, could be employed to transport the agricultural products from the fields. Articles buried in the royal tombs at Ur show that the standard of workmanship of the goldsmiths and silversmiths had by then become extremely high. The range of techniques that

became available to the Sumerians thus enabled them to create tools and orna-
ments that were different in kind to those that had been made by their hunter-
gatherer predecessors.

These various developments greatly complicated the administrative tasks of
the temple authorities. Here also new problems gave rise to new procedures.
Different people were doing different kinds of work. This had to be acknow-
ledged and paid for individually, which required the keeping of records and
the giving of receipts. This could only be done when an appropriate form of
writing and recording had been invented. Conventional signs were used for
numbers along with, at first, simplified pictures of the objects to which these
numbers referred. Later these pictures also became conventionalised. The
symbols in question were scratched, or more often impressed, on damp clay,
which was then hardened by heating. The resulting tablets provided a permanent record of transactions, and thus constituted the temple accounts. Many of
these baked clay tablets have survived among the remains of the temples, and
these have provided modern students with a wealth of information about economic and other activities in these early city-states. They represent the first
known instance of the use of writing. It seems clear that the invention and
early development of this very important device was stimulated by an urgent
practical need, namely to make possible the administration of these emerging
civilizations.

The Sumerian temple authorities also engraved life-like pictures of animals
and of human activities on cylinders composed of damp mud. The cylinders
were hardened by firing, and could then be rolled over clay that was still damp
after being used for writing; this therefore impressed the seal of temple authority on what had been written and could therefore serve as a receipt. Similarly a pot or box could be closed by means of a layer of clay on which the
mark of the cylinder had been impressed; this discouraged any unauthorized
person from tampering with the contents, since this would break the seal.
These pictures are also of interest in themselves. One, for instance, shows a
wheeled chariot, and another bound captives, so wheeled vehicles were being
used, and war was emerging as a factor in the social scene, as also was probably slavery.

The writing on the tablets indicates that by about 5000 years ago (i.e. 3000
BC) each of the small city-states had its own local god. He was regarded as the
owner of the land. The men and women had been created merely as slaves to
serve his needs. The temple authorities served him, and the king was his
agent. For the right to farm the god's land the peasants paid a tax in kind, per-

haps as grain or flax or wool, and the temple authorities gave receipts for these and took charge of them on the god's behalf. It was also the duty of the temple authorities to maintain the irrigation system, to draw up codes of law and to deal with disputes. In addition, the temple premises were used for education; here pupils were taught to read and write the difficult script, to use weights and measures, and to make calculations needed for the temple's commerce. Thus the temple authorities were largely responsible for the administration and the education, and in addition they acted as a bank, storehouse and finance corporation for their respective city-states. They were viewed, and apparently viewed themselves, as doing all this on behalf of their city's god.

The concept of a local god or spirit, which helped to maintain stability in hunter-gatherer societies, had with the increase in social complexity and prosperity apparently become transformed into a concept of a god who owned both the land and the people that comprised these states. This new version of an old concept was still an influence for social stability, for it meant that the existing social order could not be criticised or challenged. The individual was likely merely to conform to orthodoxy and to the disciplines that this imposed, for there seemed no point in using one's own mental abilities to question matters that had in any case been determined by one's local god. The individual's role, as servant of that god, was to accept the ordained role in a class society, to do what was required, and to be an effective cog in the god's (in effect the temple's) administrative system. In such a social context the natural curiosity and excellent general-purpose brain, which at earlier stages had been able to function constructively, had become seriously under-employed; these intrinsic qualities were given no opportunity to rock a boat that might easily have become unstable.

The Sumerians created a number of small separate city-states. Each had its own merchants, craftsmen, peasants and soldiers, and its seals and records, along with its own local god, its mythology, its scribes and its temple administrators. In each, its temple authorities administered these many forms of activity, acting as the nucleus of a single, more or less integrated, society or state. These various developments would have been proceeding more or less simultaneously, and would have been constantly interacting with one another. The Sumerians doubtless did not analyse or plan this complex social organization that they had in fact themselves created; they merely employed their energy and intelligence, supposedly always on behalf of the god, to each contingency as it arose. They thus created what may have been the first form of civilization to come into existence on the Earth.

The language, economic system and religion of all these various Sumerian city-states were essentially similar. Yet these states were frequently at war with one another. Here then human societies were interacting in a ruthless and destructive way. Members of the same species, *Homo sapiens*, were deliberating setting about killing one another. Other species of animals do not usually do serious harm to fellow members of their own species unless they are unduly crowded. They may contend aggressively for territory or for mates, but the loser registers defeat at a fairly early stage, and is then allowed to leave the scene without being molested further. Groups may similarly confront one another, to be followed usually by the retreat of one of them from the disputed territory. It seems that with the advent of city-states such retreats usually no longer occurred. Large concentrations of wealth now existed; doubtless the incentive to retain this was strong; also its presence would have tempted others to try and acquire it. These states were in a position to train some of their members as soldiers and provide them with the necessary weapons for war.

At about the same time as cities were emerging in Sumer a comparable series of events was taking place independently in Egypt. In this case it was the fresh silt spread widely by the annual flooding of the river Nile that was responsible for the exceedingly fertile soil. There were of course some important differences. Here also there seem at first to have been a number of small separate communities, but these became amalgamated at a very early stage, forming a single kingdom. There were no longer separate states, and hence no wars between them. This Egyptian kingdom had no organizations comparable with the temples in Sumer. However its head was the king or pharaoh, and he was regarded as divine and as acting as an intermediary between the gods and the people. He received tribute in exchange for the right to work the land. Under him were a number of nobles, each of whom acted as his deputy, collecting revenues and supervising administration in the district for which he was responsible. The pharaoh was expected to use his divine power to ensure the fertility of the fields and the welfare of his people. In practice he attended to these matters through the activities of officials who were required to take steps to predict the time of onset of the Nile's annual flood, to maintain and develop the system of irrigation, to provide armed forces and to organise merchant expeditions to import such materials as timber from north Syria and copper from Sinai. All these transactions needed to be recorded and, as in Sumer but independently, a system of writing was developed; the symbols were quite different, and they were written on papyrus made from reeds instead of on clay tablets. Thus the pharaoh, the nobles and their officials together came to perform essentially the same ideological and administrative functions as had the temple authorities in Sumer, though here in a different setting.

The peoples of both Sumer and Egypt were developing their new and complex societies in alluvial valleys that were poor in raw materials, and they had to import these in exchange for their surplus food. This brought them into contact with agricultural communities that had continued to remain simple and essentially self-sufficient. If they happened to have raw materials that were required by the new civilised world then, by exporting these, they could gradually accumulate wealth and so become secondary centres of civilization; these however were based on trade, and not on any special fertility of the soil. There was for example an agricultural village known as Assur in the pleasant foothills some distance up the river Tigris; here the Assyrians gradually developed a typical secondary centre of civilization, and in the process took over in essentials the script, the codes of law, the ways of commerce and the type of religion that had already been elaborated by the Sumerians a short distance to the south of where they lived.

Sometimes an established civilization tried to limit this type of growing power in an undeveloped area, since this would have limited its opportunities for exploitation. It would send armed forces to occupy the region, and perhaps set up an administration there. It was then occupying a region and administering a people other than its own. This was likely to be resented and to stimulate an effort to regain independence. In such a case the scales were not necessarily weighted in favour of the imperial power, for communications were poor, weapons limited in their effectiveness, and resistance might flare up at many points.

A relatively less developed country was also capable of conquering a civilised one. This could occur if the king and administration of a civilised state became inept, which could easily happen when kingship was hereditary and its power absolute. If the less developed people were to reap the full benefit of their conquest they had to conserve and learn to control the civilization they had conquered. In such a case the civilization was not greatly affected by having been conquered. Thus the Amorites, a relatively underdeveloped people, established themselves in Babylon, and under their king Hammurabi conquered the whole region of Sumer around 1800 BC. Hammurabi compared the codes of the various Sumerian city-states, correlated them and added modifications of his own. He instructed that the code resulting from this synthesis should be engraved on a shaft of stone which, characteristically, was surmounted by a sculpture portraying himself as receiving the code from the god Marduk.

Discoveries at Kultepe are also interesting. Here high in the Anatolian plateau, not far from the modern city of Ankara, lived a colony of Assyrian merchants in about 1900 BC. Apparently their settlement caught fire and they departed hurriedly, leaving their possessions behind. These still remain at the site, and it has now been possible to separate the remains of their houses, their furniture, and a great store of business documents from the overlying debris. These show that the merchants were acting as employees of the Assyrian administration, and that their task was to promote its trade with various small towns in what is now Turkey. Each of the latter, together with some country around it, still functioned as a separate principality or chiefdom. Caravans of laden donkeys travelled regularly and safely between Assur and Anatolia. The Kultepe colony was situated on the outskirts of one of the Anatolian towns, and the merchants were on excellent terms with the local people. Thus Assyria, itself a secondary centre of civilization, was already encouraging by friendly trade, in 1900 BC, the development of a tertiary centre in Anatolia. By about 1650 BC the separate Anatolian principalities had merged under a separate authority to form a Hittite kingdom.

Thus by about 1500 BC the ancient civilizations of Sumer and Egypt, with their rather formal imperial structures, were still active. There were also the newer, but rather similar, civilizations of the Assyrians and the Hittites. These four major empires were frequently at war with one another. However within the intricacies and at the peripheries of this framework there was much that did not conform to this formal pattern. There were small towns that lay too far from the imperial centres to be firmly under their influence. And there were pastoral groups who moved their herds across the more arid and inhospitable portions of the whole area, paying scant attention to state boundaries or to the conventions of civilizations.

Some consequences of this diversity were apparent by about 1200 BC The Egyptians and the Hittites had long been at rivalry. However at about this time the Egyptians were, for various reasons, forced to retire to their Nile homelands, and the Hittite empire was destroyed by another people in Anatolia. This gave the people in the intervening region, comprising approximately the modern Lebanon, Syria and Palestine, an opportunity to develop along their own lines. Only the Phoenicians and the Hebrews will be mentioned.

The Phoenicians were a commercial people who inhabited a few coastal ports. They had no elaborate state or empire. Biblos, on the eastern Mediterranean coast, had been a commercial outpost of Egypt for many centuries, and the resident Phoenicians had become skilled craftsmen and artists, well versed in the modern ways. The civilization in Crete had recently been destroyed, and this, along with the eclipse of Egypt, had left the eastern Mediterranean largely free of merchant fleets. This gap the Phoenicians now filled. They built ships, and in these they sailed to trade, perhaps on their own private initiative rather than as servants of any state. For this purpose they required for their accounts a form of writing that was flexible and easy to learn. They adopted and developed a device known as the alphabet, in which each symbol corresponded to a particular sound. The total number of these symbols, known as letters of the alphabet, was quite small. Any word in their language could be phonetically represented, or spelt, by means of these letters. Thus the written words symbolised the sound of the words used to symbolise objects or concepts. Speech (p.77-79) and writing (p.95) thus became directly interrelated. This method of writing was much simpler than earlier ones, and it opened up a whole new world. Writing ceased to be the monopoly of a small group of scribes who had learnt the art through long years of toil. It could now easily be learnt and used by anyone, and for all kinds of purposes. Technique, knowledge, experience or imaginative thought could be recorded in writing, and thus preserved and later read by anyone who cared to do so. Also the experience of a community could be recorded in writing at the time, so making it possible for its history to be known to later generations. In fact, however, the potential results that could have flowed from the invention of an alphabet were scarcely put into effect until Greek culture began to flower some seven hundred years later.

The Hebrews established themselves in Palestine. They had long been a pastoral semi-nomadic people living as small communities on the fringes of the desert. Each person would still have done a wide variety of jobs, and each would have been a respected member of his tribe rather than a cog in a wheel. In these respects they would have differed from the great civilised states that had grown up around them. These latter were highly complex and certainly not nomadic, and seasonal changes and the fertility of the soil were crucial to their well-being. Consequently the gods of these civilizations became inseparably linked with matters of this kind. The Hebrews, on the other hand, were less dependent on the characteristics of any particular area, for if necessary they could always adjust simply by altering their location. Their god was therefore not a local deity who permeated and merely formed part of local processes; on the contrary, he had no such restrictions. Sun and Moon obeyed him. Also the early civilizations taught that their gods had ordained the elaborate social systems of their people; the Hebrews had no such elaborate system, and hence no need to attribute such responsibilities to their god. They had little in the way of social classes, and hence could readily believe that their god valued equally the personality of each and every one of them.

The Hebrews thus came to believe that at a certain moment—according to tradition while they were under the leadership of Moses—their god deliberately chose them as his people. He need not have done so. The fact that he was free to choose showed that he was no ordinary god, but that he was above these gods, a god of gods, the one and only true God. This God had entered into a covenant with them; they were to remain loyal to Him, to pay attention to His teachings and to obey His laws. He, in return, would guide and protect them. All of them stood equal in His sight. This system of belief, which they had themselves created, was well adapted to meeting their current needs and circumstances. Their belief that they were God's chosen people gave to them a sense of importance and comradeship, and of history that looked forwards towards the fulfillment of the prophecies, as well as backwards.

Much later, after the exile from Babylon, and by then as Jews occupying Judea, they based their constitution on the laws of God, as interpreted by Moses and the prophets. The way these laws were applied to the details of daily life was worked out by their governing body. Popular education was encouraged, thus enabling as many people as possible to appreciate the teachings of the prophets. The Hebrews thus created a type of society that was in its time remarkable in its own right. It was also remarkable in that it was later to prepare the way for Christianity.

The achievements of the Phoenicians and the Hebrews showed that small communities could certainly be creative. A few centuries later, and also in the eastern Mediterranean area, this was demonstrated even more definitely by the Greeks.

A group of astronomers
"The School of Athens" from "The Stanza Della Segnatura" painted by Raphael

Chapter 12

THE GREEK CITY-STATES

The activities of the people known as Greeks were also very important. Most of them lived on mainland Greece or on islands in the Aegean Sea. In either case the useful land available to any one community was restricted to a small area of relatively flat ground located between the mountains and the sea. The individual communities were therefore small and isolated. The long dry summers were not suitable for growing wheat or barley, but olives and grapes grew well. The Greeks needed to trade these for grain that could be produced more satisfactorily elsewhere. To do this they had to learn to build ships and to sail the seas. The Phoenician experience was helpful here (p.100). The Greeks also adopted the Phoenician alphabet, and improved it by adding vowels. The seas, once mastered, provided a broad highway that could be used for trade, or for maintaining contact between the scattered Greek communities, or for reaching such places as Egypt where they could learn by direct experience something of the ideas and achievements of the contemporary civilizations. The Greeks were near enough to the latter to be able to draw knowledge and inspiration from them, but also distant enough to be able to develop their own independent thoughts and ways of life. Also a Greek community that had outgrown its restricted living space could use the sea to transport members to some new location, for instance on the Black Sea coast or in the south of Italy or France, where a new Greek settlement could then be established.

These various Greek communities usually built some kind of civic centre which acted as a focal point for the transaction of trade and the discussion of affairs. There was usually also a temple; this however was not particularly important, and there was no priestly caste. Their concept of their origin and place in the world was largely based on the Homeric account of the earlier Mycenean kingdom, as transmitted by word of mouth across the 'Dark Ages'. It portrayed the gods as little more than particularly well-developed men and women, and therefore had little constraining influence on the Greeks; indeed it may have tended on the contrary to encourage their self-expression. The activities of their brains were therefore much less constrained than were those of most of the other peoples at that time; their thoughts could flow more freely.

A few of these small city-states formulated fairly democratic constitutions. The largest and most important of these was Athens. Here laws, codes and ways of conduct were evolved, and for this they had no precedent to guide them. A basic distinction was made between citizens and non-citizens; the former comprised all the men who had been born of citizen parents; the latter comprised the women, the slaves, and the resident aliens. The citizens therefore remained a minority, though a large one. The policy of the Athenian city-state was determined at meetings of these citizens at the 'Assembly'. Participation was direct, every citizen over the age of eighteen being entitled to speak and vote, and resolutions passed by those assembled were deemed to represent the will of the Athenian people. Appointments to posts that involved routine administrative work, such as serving in the courts of law or on the council which prepared agenda for the Assembly, were made by lottery from among all the citizens, these appointments being held for one year only. Payment was made for these duties, but only at a rate to compensate a man for the loss of his normal earnings. By this means rich and poor could serve equally, and without either pecuniary loss or gain. One result was that over a period of time most of the ordinary citizens obtained some firsthand experience of participation in the administration of their community.

The Greeks therefore, at least to some extent, appreciated that persons were of equal potential value, and they put this concept into practice. We know now that the combination of fundamental equality and individual uniqueness in persons is based ultimately on such matters as the potentialities of genes, their reshuffling during conjugations and interactions between the resultant organisms and their environments; the Greeks knew little or nothing of such matters, but they intuitively sensed the overall result in terms of human nature. Here then, though only within the limits imposed by their idea of citizenship, the different types of resulting personalities were able to interact freely, and out of these interactions to construct policies designed to promote the common good. In a sense, but now in a very different context, this was similar to a small hunter-gatherer community, sitting at home at the entrance to some icebound cave and planning the activities of the morrow (p.79-80). The context is different because, owing to developments of various kinds since that time, the Greek communities, though still manageably small, were no longer minute, and because they had more technical and material resources, more contact with other communities and a greater understanding of their situation. In other words, the base from which they operated was now much more complex and advanced. They therefore felt more assured, and did not need a mythology that resisted any tendency to depart from established procedures. On the other hand, they also differed from for instance the earlier Sumerians and the contemporary Egyptians in not being weighed down by a relatively enormous administrative apparatus that likewise required mythologies for its maintenance. Thus in the Greece of the 6th to 4th centuries BC the excellent general-purpose brains common to humans (p.83) were able to operate without undue restraint, and to interact freely with one another. The results of this freedom and interplay could be put into effect, and if this in fact caused some rocking of the community boat (p.96), then so be it; the boat was not too unstable, or its navigators too unsure of themselves, and the democratic venture might perhaps prove beneficial to all concerned.

One result of this approach was that the Athenians of the 5th century BC came to conceive the good life as based on balanced self-development. The potentialities of a citizen should be developed to the full. No special feature should be exaggerated at the expense of others; this would lead to a lack of harmony. To live fully one should practise gymnastics and athletics, and experience the rhythms of song and dance. The mental life of a citizen should likewise be developed. He should take an intelligent interest in the views of philosophers and other citizens of special learning, and to some extent his actions should be guided by them. He should be proud and excited by the work of those creative artists who, encouraged by the prevailing atmosphere, were producing poetry, drama, sculpture and architecture of a very high order; these achievements from within his community he should feel to be in some degree his own. Such a person would naturally take a share in the activities of the community, for example in the courts of law, or in the making of decisions in the Assembly, or if necessary on the battle field. The community was small, and his participation was direct; he did not, for example, act in the Assembly merely indirectly by electing a representative. There was no reason why there should be any conflict between his personal interests and those of his community; both were intertwined, and what benefited the one should also benefit the other. Life was all too short, and he had no wish to spend time and energy accumulating an unnecessary superfluity of goods; there were more meaningful things to do.

To the Greeks the great merit of life in a Greek city-state, as opposed to that in Persia, Egypt and other non-Greek states, was that it enabled men to do things, such as gymnastics or the writing of history, simply because these things seemed worth doing for their own sake. A man's life was not necessarily confined to a doing of things as a matter of routine, merely because they needed to be done. They considered that this quality of their city-states brought out the specifically human characteristics that existed within people, and so emphasised the difference between them and animals. It thus gave scope for the humanities to flourish. The Greeks appreciated that these advantages were associated with the relatively small size of their communities. This condition, originally imposed on them by the nature of the terrain, was now seen as being positively advantageous.

This type of outlook and of social organization led to a great surge of creativity. For example, to be effective in the Assembly a speaker had to clearly articulate constructive thought based on a preconsidered viewpoint. The teaching of oratory became something of a profession, and the art of persuasion was at times unduly emphasised. Also in 5th century Athens certain individuals began to comment on the problems and vicissitudes of human life in forms that could be dramatised. Thus something altogether new, namely theatre, was developed, and presently plays with a meaningful human and social content, written by men like Aeschylus, Sophocles and Euripides, and more satirically by Aristophanes, came to be performed before huge audiences at the annual spring festival. The Athenians also tried to ascertain how individuals behaved towards one another, and to consider what was meant by such words as true, good and beautiful. Socrates made a point of questioning all kinds of people that he met in the street or marketplace and so, on the basis of patient dialogue, he tried to evolve clear definitions of various problems related to ethics. New forms of insight and participation were being generated.

History also became relevant. Some Greeks lived in Asia Minor, and thus on the fringe of the Persian Empire, and they needed to know something of the background of their overlords. Herodotus endeavoured to meet this need by travelling to the countries in question and making enquiries there, and by subsequently analyzing the results, working out chronologies and so forth. Later he used the same technique to write an account of the Persian invasions of Greece; these had taken place a generation before his time, but little existed in writing. However, by questioning survivors and those who remembered what their elders had told them, he was able to reconstruct in some detail the course of the events that had occurred. He stated that his purpose was to prevent the memory of the past being lost to later generations. Shortly afterwards Thucydides carried this approach further. He wrote about the Peloponnesian war, of which he was a contemporary and in which he had played some part.

He endeavoured to report factually, to confine his attention to a narrow field comprising the war and its associated politics, to analyse the results, and thus to illustrate human endeavour as seen in this one particular context. Such studies led people to think in terms of changes that take place at the social level, so modifying the form of human society as time progressed. It was, therefore, beginning to become possible for a community to view its present condition as arising from its own actual past, instead of having to rely on mythical versions of that past. History, first made potentially possible by the development of writing (p.95), was now coming into existence.

There were also some Greeks who now gave serious thought to the nature of the natural world around them. The first such person of whom we know was a merchant called Thales who lived in the 6th century BC in the busy commercial town of Miletus in Asia Minor, and thus in a peripheral zone within the Persian Empire. In such civilizations it was still assumed that the world was run by divine beings, the gods, who were embodied in the Sun, winds and so forth. Thales however was well aware that it was men like himself, not gods, who controlled the affairs of their city, and he saw no reason to suppose that gods controlled the natural world either. No longer bound by this constraint, his natural curiosity (p.74) was free to roam. He was conscious of interrelationships between things and assumed intuitively that there must be some underlying unity that was responsible for the overall cohesion that was manifest everywhere. Presumably as a result of thinking along these lines he suggested that water was the primary substance from which all other things were made. It seems that he was led to this suggestion by observing that land could be laid down from water, as in the Nile delta, that food and seeds both contain water, and that water can exist as a solid, a liquid or a vapour. Water therefore plays some part in many important natural processes. Thus, as in science in our own day, he was trying to use a single non-mythical concept to explain a wide range of phenomena. This type of generalization was new. Furthermore, since it made no appeal to divine beings, it could be rationally criticised and alternative proposals could be made.

Anaxamines, also of Miletus, chose air as an alternative primary substance. Air, as he saw it, gives rise to wind, and to cloud, water and rain; air is also essential to living beings, and for him the breath of life was scarcely distinguishable from the body or soul. He suggested that all types of things were produced as a result of what he called the densification and rarefication of air. Here another concept familiar to science was being expressed, namely that quantitative changes can lead to changes in quality.

Somewhat later Heraclitus, another Greek who lived in Asia Minor, took the impermanence of things as his starting point. Everything was in a state of flux. One cannot step into the same river twice; a river is still there, yet the river that was there at first has flowed by and passed away. He and his followers distinguished between thought and the things thought about, and suggested that in the same way as people can, by thought, keep their own affairs in order, so thought permeated the Universe and prevented the universal flux from resulting in universal chaos.

Another Greek natural philosopher, Pythagoras, is said to have used pebbles on the beach near Croton to investigate the way in which units of matter could be arranged in space to form patterns. A study of numbers, of geometry, of units extended in space, is essential to an understanding of the various types of entity that comprise and give character to our world. Because the Pythagoreans used pebbles, not abstract points, they appreciated that number was one key to the understanding of the Universe. They were impressed by the fact that all patterns of a particular type have certain features in common. Thus the sum of the internal angles of any triangle amounts to two right angles. Such a relationship is absolutely constant; it is maintained in all circumstances, despite every flux and change. The Pythagoreans were in some respects a mystical group; they believed that the mind or soul could live an existence separate from that of the body and could live on after death. The more they could learn about absolute and eternal truths, such as those revealed by mathematics, the better it would be for their souls.

There was also an atomic theory. It was founded by Leucippus in the middle of the 5th century BC, and was later further developed by his pupil Democritus. They started from the observation that no thing can be created from nothing. They argued from this that the ultimate units of matter must be finite indivisible particles that they called atoms. Their existence could be inferred from everyday experience; for example, wet clothes hung on a line become dry, but one does not see the water leaving them; this is because it goes particle by particle, and the individual particles are too small to be seen. In this kind of way they made inferences about the unseen from the seen.

Greek natural philosophers therefore observed natural happenings in the world around them and made tentative deductions regarding their significance. One can see now, by hindsight, that their next step might have been to devise and undertake suitable experiments that would have tested the validity of these deductions. However they seem to have done little in the way of testing, perhaps because most of them were sufficiently well-to-do for women and slaves to have done all routine jobs that needed to be done by hand. Setting up experiments would have required the use of their own hands, to which they would have not been accustomed, and which would in any case have been regarded as demeaning. These natural philosophers may therefore have found it easier and more comfortable to confine themselves to speculation. If practical tasks and productive activities had been viewed with more sympathy the outcome might have been different. Theories might then not have remained isolated, their validity unchecked by tests. Ideas and experiments could have interacted, each illuminating the other, so bringing an increased understanding of both natural and productive phenomena. The birth of a systematic ongoing form of science might then have been brought forward by some two thousand years.

The consequences of this Greek approach to the nature of things were exemplified by Plato, and soon afterwards by Aristotle. The views of both are relevant, for they have subsequently exerted important influence on the underlying climate of thought. Plato had in his youth been acquainted with Socrates, and he had also been much influenced both by the mathematics of the Pythagoreans and by their concepts of the salvation of the soul. He was a member of a rich aristocratic Athenian family, and he purchased some property in an outlying portion of the city. Here, isolated from the confusion of the marketplace, he set up the first institute of higher learning, which became known as the Academy. He appreciated that thought is something different from the data received by the senses, memory and so forth. It was at this point that Plato's acceptance of the Pythagorean concept of the soul became important. He supposed that the mind or soul which created this mental life was immortal, and so could exist independently of the body. He extended this concept by supposing that similarly an 'Idea' of a thing existed and was independent of representatives of the thing in question. For example, innumerable circles could be drawn, all of which would be more or less imperfect. But there exists also the Idea of a circle, which is a perfect circle, to which all circles that are actually drawn are mere approximations. To Plato the Idea of a circle was not just a concept of the mind; it was something that, like the soul, existed quite apart from people's thoughts about it. Similarly there existed the Idea of a horse; the various actual horses were imperfect approximations to this ideal horse. Again, there was an absolute Good, which was the Idea of goodness, an absolute Truth, and an absolute Beauty. The chief duty of the soul was to comprehend these Ideas. The Idea was the thing in its perfection, and the soul was striving to understand the ideal. To look too closely at the actual world was inimical to this; the mind became confused by a multitude of imperfect circles and horses. The comprehension of an idea should be attempted by reason alone, not through the senses. His philosophy was therefore antagonistic to observation and experiment. The tendency of natural philosophers to speculate, but not to experiment, when trying to understand the nature of the world, thus became rationalised by Plato's idealistic philosophy.

The views developed by Aristotle also had a longterm influence. His father was physician to the king of Macedonia, so in his youth Aristotle probably had firsthand experience of bandages, fractures and the care of the human body. He would have appreciated that to do things with one's hands was not degrading. At the age of seventeen he was sent to Plato's Academy in Athens, where he seems to have accepted much of the teaching. Gradually, however, he became dissatisfied with the concept of Ideas. Efforts to understand Ideas that were supposed to exist apart from things prevented the proper observation and study of the actual things themselves. For Plato's Ideas Aristotle substituted 'Forms'. The 'Form' of a thing was its essential inherent nature. But for Aristotle the nature of a thing, or its Form, did not exist apart from the thing itself. Observations about things could therefore contribute to the understanding of their 'Form'. From a study of horses one could hope to learn something about the 'Form' common to all horses. In addition, however, Aristotle believed that the 'Form' of things involved not only their essential inherent nature, but also included the purpose that Mind or Providence had had in view when it made the 'Form' of the kind it was. To him the problem of the meaning behind 'Form' was more fundamental than understanding the 'Form' itself. Consequently he tended to ask himself why things are as they are, rather than how they have come to be as they are. His whole outlook was what we now call teleological.

Through introducing this concept of 'Form' Aristotle recovered the possibility of study based on observation, and he used it to good purpose in his comparative study of animals. He and his school also prepared comparative accounts of the constitutions of numerous Greek city-states, and discussed such matters as how revolutions can be avoided by the use of moderation and making timely adjustments. His work was thus concerned with a vast range of human experience, extending from the physical world, where he largely followed Plato, to biology and to social organization. This synthesis came to have an immense influence during the next two thousand years.

Between the 6th and 4th centuries BC the Greeks thus showed that to be small could help societies to be creative. To this extent they resembled the

Phoenicians and the Hebrews, but were quite different from the very large imperial states. They demonstrated that in their own kind of setting it was possible for people to free themselves from earlier mythological preconceptions, and also that societies could be run democratically, and in ways that benefited both the individual members and the community of citizen as a whole. They thus created exceptional scope for living life to the full. They used these opportunities in many different ways. They thus demonstrated that there existed in human beings a vast creative potential most of which had previously remained untapped. All this was new.

However these city-states also demonstrated that small can be weak. They remained quarrelsome and in some respects very ineffective. They were fortunate, thanks to Athenian sea power, to be able to ward off the massive Persian invasions of 490 and 480 BC However about one hundred and fifty years later Philip of Macedon overran much of Greece, which virtually ended the independence of the Greek city-states. Nevertheless these small communities of Greeks had had sufficient time, opportunity and zest for life to be able to create experiences whose memory has in some degree inspired all later generations.

Soon after Philip had overrun Greece his son, Alexander, increased the conquered area so greatly that by the time of his premature death in 324 BC a vast area that included Greece, Asia Minor, Egypt, the Near East and the northwest corner of the Indian subcontinent was being administered by Greeks or Macedonians. The result was that the laws, festivals and general ways of life came to have a certain similarity, and also trade flourished, throughout this whole vast region. A simplified form of Greek provided a common language in which people of any race could communicate with one another. This extension of Greek influence gave rise to what came to be known as Hellenistic civilization.

This situation confronted the Greek followers of Alexander with serious administrative problems. It was essential for them to retain the elaborate civilizations and the long traditions of the states they had conquered, not least because there were in any case far too few Greeks to administer them unaided. The result, for instance in the case of Egypt, was that a Greek named Ptolemy took charge of a population consisting mainly of serfs. He was accepted by the Egyptians as the pharaoh and, at least by the time of his son Ptolemy II, the new pharaohs were regarded as divine. The whole system of economy that had been developed by earlier pharaohs was in principle retained.

All this necessarily affected the outlook of the individual Greek citizen, though the way in which it did so depended on his position in the social framework. In this Hellenistic world it was a time when many an individual of whatever race appreciated the importance of understanding, and thus being able to make use of, the wide range of knowledge that had been accumulated. Hence for the natural philosopher type of Greek the new environment was stimulating. Greeks now controlled the Egyptian manufacture of papyrus, and the employment of educated slaves enabled manuscripts to be copied more rapidly than before. A common language and freedom to travel greatly helped the spread of knowledge. Alexander, who as a youth had had Aristotle as a tutor, seems to have appreciated the importance to administration of a knowledge of such matters as history, geography and engineering. The generals who followed him usually set up libraries in their capital cities. The most important of these was the Museum at Alexandria; here the Ptolemies used a small portion of the wealth derived from taxation in Egypt to assemble an enormous library, and to provide an observatory, zoological and botanical gardens, and accommodation for research work; they also paid salaries to about one hundred men of learning.

In the time of Ptolemy II the library at Alexandria is said to have contained about 700,000 rolls. The chief librarian and his staff worked on this material; they arranged the rolls in a definite order according to their subject matter, prepared a catalogue, and compared, criticised and edited the texts. This brought them up against the problems of language and its structure, which led to a recognition of the various parts of speech and of what we now call grammar. This was very helpful at a time when many people of different races and languages wished to express themselves clearly and accurately in Greek. This work on the manuscripts at Alexandria made possible a knowledge of the known. It enabled specialists to examine and give thought to all previous work that had been done in a particular field, and to present it in a logical and systematic form which the mind could readily grasp. This was an essential preliminary to further research. Thus a characteristic feature of the work of the Museum was the production of textbooks, of which that by Euclid on geometry is now the best known. The resulting increase in background knowledge, and thus in education, led to the production of more efficient calendars, maps, harbours and war machines. In such ways the work of natural philosophers directly helped the administrations that financed them. The tradition thus established is with us now. Any research needs to be based on a comprehension of the work already done, and an understanding of the concepts that are currently employed; here scholarship blended with research, and history with science.

During the 3rd century BC natural philosophers had exceptional opportunities for working without financial worry, for integrating their activities, and for developing a historical perspective. A number of important advances were made, of which two are noted. First, a Greek named Eratosthenes successfully applied the principles of geometry to calculating the size of the Earth. He knew that at Syene, the modern Aswan, the Sun was directly overhead at noon at the time of the summer solstice. At Alexandria, which was due north of Syene, the comparable angle of the Sun at this time could be obtained by measuring the relative length of the shadow cast by a vertical rod at noon. The distance between these two places was known, so it was easy, assuming the distance of the Sun was relatively-speaking infinite, to calculate the radius of the Earth. He obtained a result very close to the one that we now know to be correct.

The second instance refers to the Greek Archimedes. He worked out the principles applicable to levers. He also invented a system of pulleys by means of which great weights could be lifted. He defined specific gravity, discovered methods of measuring it and established the principles of hydrostatics. His work therefore had important practical implications, and seems to have developed through the integration of theory, experiment and practice. In these respects he was exceptional.

For a more typical educated Greek the position was rather different. In the past he had lived within the confines of a small city-state; his personal interests were linked with those of his community, and its affairs had received his personal attention. But after Alexander the small had been replaced by the very large, and he could no longer hope to exert any influence on these huge new states, especially as they were in most cases administered by autocratic monarchs. His viewpoint therefore changed, becoming centred more on personal interests, and if he had the ability and the inclination he would try to work out afresh his purpose in life and his relation to the greater world around him. The principal result was the philosophy known as Stoicism. It was essentially a liberal philosophy with a strong sense of ethics. It was not revolutionary, for it did not advocate any basic change in the general pattern of society; a slave, for example, would remain a slave, unless perhaps his master could be persuaded to free him. By encouraging people to develop a sense of conscience and duty it helped them to become self-critical. It also encouraged them to face their problems fearlessly, contentedly, and with a good conscience. It involved a God who was Himself part of the Universe and of Nature, as also were humans, who were therefore also in their own small way divine. The male person, being a part of nature, was similarly in his own small way divine. It was a lofty philosophy that brought within its compass men, society and the Universe.

The outlook of the less educated Greek was also affected. The Stoic philosophy passed him by, being too complicated. Earlier, while in Greece, the larger-than-human type of gods that dwelt on Mount Olympus would have been sufficient for his needs. However now, when settled in some distant and very different kind of city, perhaps on the banks of the Nile, these gods soon came to seem meaningless. This left him in a vacuum as regards religion, and he therefore became susceptible to the ideas of the people in whose country he now lived. In most cases these were broadly similar to those of the Sumerians. Thus although Greek ideas were taken further to the east, eastern ideas were also absorbed by many Greeks, and before long it was the second trend that came to dominate the cultural scene.

In general the basic limitations noted in Greek society remained in operation, though now functioning in a different setting. Labour was abundant, so there was no initiative for men of learning to try to invent devices that would have lessened for others the toil associated with the processes of daily living. Even water-wheels, which were already known and could have provided a source of power, remained virtually unused. Hence natural philosophy and its applications, though encouraged by governments, remained divorced from day-to-day productivity, and so did not become assimilated into the patterns of communal life. As already noted, natual history helped governments with large-scale projects such as constructing harbours and war-machines, and with their financial backing it flourished during the 3rd century BC. Nevertheless it rested on precarious foundations, and by the 2nd century concepts associated with primitive religions had come to he almost universally accepted. In the resulting climate of thought any objective consideration of human, social or natural matters once more became virtually impossible. By then the Greek initiative had therefore been almost totally eclipsed, and it was to remain in that state for many subsequent centuries.

Chapter 13

THE DEVELOPMENT OF THE "WEST"

This chapter is concerned with the portion of the world known loosely as the 'West'. The justification for special emphasis on this area is that, although for many centuries it was a remarkably backward region, yet more recently it has been principally responsible for the condition of our home the Earth being as it is today. The West is therefore very important to all of us, whether Western or non-Western, at the present time.

We can start by noting that early in the 5th century BC, and therefore at a time when the Greeks were already evolving drama, philosophy and the study of history, the Romans were one of several small communities who lived by cultivating the fertile coastal plain on the west side of central Italy; they were therefore far from the main centres of civilisation. Populations tended to increase, but fertile land did not, and these local communities were frequently at strife in their endeavour to obtain each for themselves a maximum share of this limited resource. The early Romans were therefore farmers, but in addition they were prepared at any moment to defend, or to extend, the lands they farmed. Their survival as a community was achieved through their developing a sense of internal solidarity and a capacity to coordinate their actions. Thus the Romans became a tough, practical agricultural people, with a good communal spirit, a sense of authority and responsibility, and an ability to organise. They had little interest in ideas and theories, and none of the Greek genius for exploring new fields of human experience.

The success of the Romans, as compared with their neighbours, was largely due to the policy that they adopted towards those they had conquered. They did not kill, destroy or enslave wholesale, as was the normal practice on the Italian peninsula. Instead they tried to reach an understanding with a defeated community, to leave its towns intact, and to grant to its people certain privileges. The Romans took some of the land of such people and cultivated it themselves, and they also built roads and established military settlements. They guaranteed security for the vanquished region, which in return had to contribute soldiers for the Roman legions. By such means the Romans tried to ensure that when they had been successful in a war they were able to consolidate their gains. The Roman community thus gradually increased its power and influence, and by about 290 BC the whole of central Italy lay within its orbit. This expansion brought Rome into direct contact with the Greek settlements in southern Italy. The Romans conquered these, and attempted likewise to assimilate them. However the empirical outlook of the Romans was not well suited to appreciating the refinements of Greek ways, so assimilation was only partial and resulted in the creation in this region of a mixed, or Graeco-Roman, type of society.

Later the Romans eventually defeated the Phoenicians during the long Carthaginian wars. Still later, as a result of numerous further conquests, they greatly extended the area of what had by then become the Roman Empire. This had expanded sufficiently to include parts of Britain, of north Africa and of the Near East. However these procedures proved counterproductive, for presently the Romans, partly as a result of many wars, found themselves faced with grave financial difficulties, with a super-abundance of slaves, with all-pervading internal corruption, with rebellions from people they had conquered and resistance from those still unconquered. As compared with their history some hundreds of years earlier almost every aspect of their situation and their policy had changed. The important overall result was that in the course of the 5th century AD the whole of the western portion of the Roman Empire totally collapsed.

The birth, life and death of Jesus of Nazareth took place in a part of the Near East that was under Roman control. His fellow Jews mostly regarded him as a further important Hebrew prophet. On the other hand some local non-Jews came to believe that he was the Son of God, and that he had ascended to Heaven after he had been crucified. This belief became the basis of a new religion, namely Christianity. It differed from most contemporary religions in that it endeavoured to bring hope and consolation to the poor, and this, in the prevailing economic and other circumstances of the period, was much needed. Thus Christianity, starting from a small focus, spread fairly rapidly and soon came to be accepted by many people in the Roman world.

This new doctrine was spreading among people who were in some cases familiar with Plato's beliefs (p.109), and there was much interchange between the two sets of ideas. After about three centuries the Neoplatonic form of Christianity that resulted from this synthesis differed considerably from the form in which Christianity had originated soon after the death of Jesus. As the Christian movement continued to grow it developed its own administrative

structure and tended to become something of a state within the Roman empire, and as such was resented and persecuted by the Roman authorities. However eventually, having failed to suppress it, they decided instead to join it,
and Christianity became established as the official Roman religion.

The collapse of Rome in the 5th century AD involved only the western,
Rome-based portion of the Empire; the eastern portion, based on Byzantium,
continued to function. So also, naturally, did the various states and empires
that had grown up further afield over the centuries, for instance in India and
China. However in the western area the whole Roman state apparatus disintegrated. Communal life in the 'West' therefore necessarily assumed a local
character. Small units once more came into being. Each local community had
to become as far as possible self-sufficient, producing the commodities it
needed and arranging for its own defence. It would come to have its own local
noble or feudal lord, to whom the people would provide some armed support
and other services in return for receiving protection. Thus the initial confusion
that followed the collapse of this portion of the Roman Empire was gradually
replaced by a number of local feudal communities in each of which the rights
and duties of the various members had become determined by common usage.
As time progressed power tended to become concentrated into a smaller number of larger units due to one noble dominating or displacing others in the
neighbourhood. This could result in the formation of small kingdoms.

The Roman state collapsed, but the administrative apparatus of the Christian Church was largely able to survive. The Church therefore exerted some
civilising influence during the turbulent period that followed the fall of Rome.
It provided the main link of continuity, and it was through this medium that a
certain very small portion of the intellectual resources developed by the classical world was transmitted to early medieval society. The concepts in question
had resulted from mutual interactions between Christian, Platonic and Stoic
beliefs. The mystical core was belief in the Virgin Birth, the Resurrection and
the Holy Trinity. The present temporal life was merely a preparation for the
life to come, and should be lived in a way conducive to the love of God. It was
important that people should be neighbourly, and that they should appreciate
the community aspects of their lives and the responsibilities that these entailed. To obtain interest from loans was, for example, to receive gain without
having undertaken any corresponding labour that would justify it, and was
therefore unethical. This general attitude harmonised well with the complex
feudal society that was developing, where each unit functioned through collec-

tive effort, and in which every individual had responsibilities to those both above him and below him in the social hierarchy.

There was in Christian thought a tendency towards the renunciation of the flesh in favour of a more ascetic way of life. A group of people might wish to withdraw collectively from the daily tumult of the world in order to devote themselves to prayer and meditation. But to live thus simply required initial preparation of land and the construction of suitable buildings. During the 4th to 6th century AD a number of these institutions, known as monasteries, were established. The monks had to cultivate the land, to tend the fish ponds, and to maintain the buildings and their furnishings. They also had to participate in an elaborate programme of religious services, and at precise times. They had to calculate the dates of the religious festivals, and to copy and edit various texts related to their calling. To perform these tasks satisfactorily required adequate training, and the monasteries provided the tranquility and the general frame of mind in which such training could be given. They soon became the chief centres of learning, and hence also of education. These active centres of religious life, scattered through the feudal countryside, demonstrated what Christianity could at its best be like. Interplay between laity and monastery was close. Bequests were frequent. Thus the monasteries gradually became increasingly influential. By their example they encouraged orderly effort, technical efficiency and careful scholarship. They became very important institutions in the early Western medieval world.

The quality of these monastic activities was further improved when in the 10th century the scholar Gerbert of Rheims undertook a careful study of a Latin translation of Aristortle's work on logic. The translation had been made, with commentaries, by Boethius during the period when Roman civilisation was collapsing, and it had been largely ignored since that time. The outcome was something of a revelation. It was found that Aristotle had been able to classify the ways in which an object can be thought about into categories, such as quality, quantity, space, time and so forth. Types of valid argument were discussed, and also sources of error and deception. Aristotle, it seemed, had brought order out of chaos in the whole perplexing field of human thought. Further, these methods of logical analysis could be applied to the difficulties encountered in many texts, to the editing of manuscripts, to theological argument and to a consideration of Plato's 'Ideas'. The scope for scholasticism was thus extended and rational analysis was applied to its content. This emphasis on rationality was not regarded as inimical to religion; on the contrary, man was using his reason, given by God, to try to reach a better understanding of matters pertaining to God. It was a static approach rather than a dynamic one, and it was applied to ideas rather than to things, but it did lead intellectual activity away from the monotonous uniformity into which it had congealed in late classical times. It made people think about why they thought what they thought, with the result that they became more mentally alert.

Towards the end of this same 10th century conditions in the West were becoming more settled. The nobles, as a result of squabbling among themselves and sometimes with their king, had gradually established boundaries for their estates which, backed by their castles, were not easily shifted. Further conflict was unlikely to yield positive results, and was less often undertaken. It followed that by the 11th century kings were more able to collect taxes with which to finance the administration of their realms. For example, soon after William of Normandy had conquered England he appointed sheriffs to each district whose duties included the collection of taxes and the administration of justice in their areas. To prevent tax evasion William instructed that a list be drawn up, known as the 'Domesday Book', which recorded the property and economic status of every person of means. The preparation of this record required the employment of clerks who could read and write. To use it to assess appropriate tax demands also required literacy, as also did the receipt of taxes and the payment of the sheriffs and their personnel. The sheriffs were able to discharge their duties efficiently only if they had some knowledge of logic. The growing administration, itself only possible on account of the more settled conditions, thus created excellent opportunities for persons who had been trained in the scholastic procedures that had been so recently elaborated. Education therefore prospered.

These more settled conditions also promoted changes of another kind. Human-power, so plentiful in classical times, was now scarce. The general situation encouraged people to try out innovations that might increase production. Horses were harnessed by a method which did not constrict the windpipe when they were drawing a load, and they could then pull a heavy plough through the ground more efficiently than oxen. Watermills, known since classical times but rarely used, were now built in large numbers and used for grinding corn; there were about five thousand of them in England alone during the 11th century. Use of horse-power and water-power eased the strain on human muscles and gave people more time and energy to widen the range of their activities. The potential value of technology in circumstances when man-power was scarce was beginning to be appreciated.

The resulting increases in productivity sometimes led to a surplus of particular types of goods, and hence to more local trade. A major share of the benefit often passed to the feudal aristocracy, and they became accustomed to exchanging some of it for silk, spices and other luxury goods obtainable from the East. Each summer during the 11th century caravans moved southwards through Europe, often travelling along the remains of Roman roads. At various points they received woollen and linen goods and other articles for the export trade. They struggled over the Alpine passes, and so down to Padua. Consignments of goods were likewise brought from the East, disembarked at Venice, and thus also reached Padua. There the two sets of merchants bargained for the exchange of their respective goods.

Trade, whether within Europe or with the East, had secondary consequences. More or less permanent colonies of merchants were soon living at the reception points for goods; they needed food and services, and this led to more people living there. Thus new towns arose, and the remains of old ones revived. Also it became worth specialising on the production of particular types of material suitable for trade. The most important of these was cloth. Sheep were reared especially for their wool, which was converted into cloth at various centres of manufacture, and these grew into industrial towns such as Bruges and Ghent.

These new towns, whether commercial or industrial, had no special ties with feudal society. Two different and more or less independent kinds of community were therefore coming into existence. There was feudal society, now relatively stabilised and conservative, which was concerned with the land and its productivity; this involved the great majority of the population. Here the life of the individual person was circumscribed by his duties. However this was often regarded as an asset rather than a burden; it reduced the need to make decisions and so, within the prescribed limits, it left an individual free and tranquil. One knew that one was playing a well-defined role in the activities of one's community; one belonged to it, and was secure within it. It also had the merit of fitting comfortably within the contemporary Christian setting. By contrast, the new urban society of burgers and bourgeosie, still very plastic, was primarily concerned with the purchase of raw materials, their conversion into goods, and the sale of these for profit. Here life was therefore primarily industrial or commercial. There were no feudal ties, and no close harmony with Christian ethics; an individual was free to scramble as he liked for profits, perhaps becoming rich in the process or perhaps falling by the wayside. The individual did not have the same sense of security, or of being an integral part of the community in which he or she lived.

Industry and urban life rapidly became more important during the 12th and 13th centuries, and their effect on feudal society therefore also increased. The existence of a town in the vicinity gave to a serf a place into which he could disappear and obtain work, and so escape from feudal ties. His overlord might be glad to forestall this possibility by accepting a cash payment to end his servitude. Alternatively the serf might be allowed to pay a rental, instead of giving services, for the land he had been cultivating. As the urban form of society developed it therefore gave rise to pressures that tended to disrupt the feudal form. In consequence this latter, with its services and its social hierarchy, was being slowly replaced by a form of society that was based on a monetary system.

The scene in the West was also influenced by events that took place far outside its area. These began in the 7th century AD when, in a small Arabian trading town called Mecca, a certain Arab became very conscious of the relative backwardness of his people, who still worshipped petty gods, practised infacticide and were continually engaged in feuds. He was also aware, by contrast, of the nature of the Jewish and Christian faiths, and he came to believe that he was at times in direct communication with God, and that He had entrusted to him the mission of bringing the Arabs to conform to appropriate ethical standards. He regarded Jesus as a major prophet, and he considered that he himself had a similar status.

The Moslem religion that Mohammed thus initiated seems to have met a widespread social and spiritual need, and not only in Arabia. It spread rapidly, and soon became the dominant religion in an area that stretched from the Indus in India through Persia (Iran), Syria, Iraq, Arabia and Egypt, and thence along the whole north coast of Africa, together with extensions across the Mediterranean that included Sicily, part of southern Italy and almost the whole of Spain. Throughout this whole area agriculture, sometimes aided by irrigation, flourished. Trade also flourished. New cities were built. This Moslem world, extending from the Pyrenees to the Indus, was linked by the waters of the Mediterranean, the Red Sea and the Indian Ocean, on all of which Moslem ships sailed about their business; indeed they travelled much further, to trade with other parts of India, with the East Indies, and with China.

In Persia the Arabs became aware of the presence of early Greek manuscripts that had been translated into Syriac. They had these translated into Arabic. This introduction to the works of Aristotle, Euclid, Ptolemy, Hippocrates and other Greeks opened to the Moslems a whole new world of experience, and they gradually related it to their own contemporary scene. After assimilating its content they began to make their own contribution to background knowledge. They made improved astronomical instruments, and followed Ptolemy by bringing astronomy to bear once more on problems of geography and navigation. They introduced the system of symbols for numbers that they had found the Hindus to be using in India; this greatly simplified mathematical calculations, as compared with the use of Graeco-Roman notation. They extended the earlier Greek study of the refraction of light, applied it to the use of lenses and to their own observations of the anatomy of the eye, and so were able to construct spectacles that corrected defects of vision. They were interested in drugs used for medicine, and they established excellent hospitals. In general, they were more concerned with experiments and their relation to practical utility than had been the early Greeks whose works they had brought into active use once more.

Later, during the 11th and 12th centuries, Moslem power was on the wane, whereas that of the West was relatively in the ascendant. The Moslems withdrew from the south of Italy, from Sicily and from parts of Spain. The Christians who moved in after they had left soon realised that these Moslems had experience of a great body of learning that they had acquired from the early Greeks, and of which the Christian West was almost wholly ignorant. It was not easy for scholars in the West to comprehend the Arab manuscipts. They were written in a difficult language, and were concerned with facts, ideas and modes of thought with which the Western tradition was not familiar. But it was exciting, for no one knew what great discovery might be revealed by the translation of the next document. Scholars from many parts of Europe were drawn to the regions of Moslem retreat. Thus during the 11h century a small stream of Arab learning was reaching the West, but in the course of the 12th century it became a flood. Before the end of that century translations from Arabic to Latin had been made of: (a) The whole of Euclid's geometry: (b) Ptolemy's great integrated work on astronomy, which totally supplanted the odd fragments of astronomical knowledge that had been available in the West: (c) Greek and Arab researches on light, on vision and on the eye, a field that was wholly new: (d) The medical writings of Hippocrates and Galen, together with some Arab contributions: (e) Many of the works of Aristotle, including much of his physics and biology. These various works had to be correlated, and the resulting matter presented in a form that was relevant to the current outlook and interests of the West. Such tasks were well suited to the abilities of Western students who had by then been trained in scholastic logic.

The growth of education in the West was giving rise to increasing numbers of scholars, and later in life many of them became important and influential people who were involved in administrative and commercial state activities. The influx of great blocks of Arab learning that revolutionised whole fields of knowledge excited the enthusiasm of people of this type. In Paris Abelard, one of the great teachers of the period, became impatient with the prevailing emphasis on Plato's 'Ideas' (p.109); he therefore emphasised that whatever the nature of an ideal rose might be there remained visible differences between individual roses, and that these also should receive attention. Another scholar, Adelard of Bath, suggested that although it was God's will that plants should spring from the soil, yet the way in which this took place was "not without natural reason too". In other words, God wished that plants should grow, but the way in which He put this into effect was by means of natural processes. Thus, partly under the stimulus afforded by Arab experience, Abelard and Adelard developed a different conception of the relation between things, people and God; it encouraged the study of things and processes in their own right, and this in turn facilitated the understanding of the concepts that were being received from the Arabs.

In such fields as geometry, chemistry, optics and medicine the earlier Greek and Arab work was comparatively easy to assimilate. But the work of Aristotle, known in its entirety by about the end of the 12th century, presented a more difficult problem. It gave a comprehensive and closely integrated account of the whole Universe, which it interpreted in terms of natural causes. The Christian religion had its own different account, based in this case on revelation. The Church endeavoured at first to treat this challenge to orthodox Christian doctrine as a heresy, and the teaching of Aristotle was banned in 1210. But the enthusiasm engendered by the discovery of a comprehensive new philosophy was so great that it could not be suppressed. Two Dominicans, Albertus Magnus and his pupil Thomas Acquinas, accordingly developed instead a synthesis or compromise between the two systems of thought. One result was to establish a separation between faith and reason; faith was an appropriate guide to such matters as religion and morality, but for the understanding of the natural world reason was more suitable. Thus theology was separated from philosophy; each had its appropriate sphere of action; both were guides towards a wider truth. This solution was accepted by the Church.

Thus before the end of the 13th century the works of Aristotle had been accepted in the West and were forming an important part of the schedules in the universities that were coming into being. One result was that philosophy

CREATIVE LEAPS THAT SHAPED THE WORLD

had been freed from theological entanglement, and it could therefore pursue its own independent enquiry in the field of natural philosophy. Another was that the West now had access to Aristotle's comprehensive theory concerning the natural world. It proved not to be a good theory, but at least it was a theory, and the only one available.

The mutual interactions between productivity, trade and education continued to operate during the 14th and 15th centuries, slavery had been virtually eliminated, and there was always a shortage of humanpower; this was periodically aggravated by devastating outbreaks of bubonic plague. Any improvements in methods of production were therefore likely to be welcomed. Manual workers engaged in productive activity occasionally brought to light useful departures from normal routine. Such men were usually illiterate. However students at universities were given some groundings in the principles of agriculture, navigation, medicine, the manufacture of cloth and so forth. Such men often came to hold important executive posts and they were then well placed to promote the utilisation of constructive observations made by working men. Both types of person could therefore cooperate in the improvement of techniques by contributing their different, but complementary, experience and ability.

One important innovation was that by the 14th century watermills were being used for a number of purposes in addition to grinding corn. They were employed to drive the bellows used to heat furnaces; previously iron could be heated only to a bloom, and this had then to be hammered into the approximate shape required. The extra heating resulting from the use of mechanically driven bellows was sufficient to melt the iron, which could then be cast directly and precisely into the form that was desired. At about the same time waterpower was being used to drive the grindstones used to make metal edgetools. People were becoming power-conscious and machine-conscious.

Many of these applications involved the use of cogs. With a falling weight as a source of power, an escapement mechanism and appropriate cogs it was possible to construct a mechanical clock. Clocks were the first all-metal machines. They were also the first precision instruments, and clockmakers became pioneers in the field of precision engineering. Clocks were therefore important for technology on account of the training that they gave to those who constructed them. They were also important in their own right. By about 1500 most towns had a clock set up in some public place. They were not particularly accurate that had to await the incorporation of a pendulum mechanism but they served to make people time-conscious and enabled them to coordinate their activities more effectively than before.

The reproduction of a pattern of printing from a wooden block had long been known, and by the late 14th century metal blocks were occasionally used. By about 1450 satisfactory methods of casting type were being developed; hundreds of metal blocks of each letter, of similar size and interchangeable, could now be made, and these could be assembled to form a page of writing. Once the type had been set an unlimited number of copies could be produced. The slow work of the early copyists was eliminated, and so were the inevitable errors that this copying procedure had entailed; by contrast, every copy of a printed edition was necessarily identical. The new process led in the following century to a great demand for books, and large numbers were printed; some were derived from manuscripts that had long been in existence, such as the Bible and the works of the early Greeks, and some were by contemporary authors. The multiplication of books stimulated the growth of literacy. By means of printed pamphlets it was relatively easy to disseminate unorthodox views; orthodoxy therefore no longer had the same monopoly of thought. People became better informed and more mentally alert.

Seagoing ships were also improved. Wind, which provided power, could blow in many directions. Changes in sails and in the type of rudder enabled ships to tack, and so to pursue a course that was, overall, more or less straight in the face of contrary winds. A compass needle mounted on a point, and so able to move independently of the ship's motion, was in use by the end of the 13th century. These important innovations enabled ships to be held safely on their course even when out of sight of land. At this time trade with the East by way of the Black Sea had become impossible, and it was difficult via Syria or Palestine due to the local situation in these areas. The Italian cities compensated for this loss by increasing their trade within Europe. This was possible because more land was now being cultivated in Europe, and the output per acre was larger; also manufactured goods were being produced in quantity. This change in trade relations was facilitated by the new navigational techniques, for ships could now leave the shelter of the Mediterranean at Gibralter and sail reasonably safely through Atlantic waters to reach, for example, Bruges, Antwerp, London or Hamburg. Such cities became large centres of banking and commerce. The Italian cities therefore no longer remained wholly at the centre of the European trading complex.

This increase in commercial and industrial activity in Europe further increased the strain on feudal society, and by the 15th century it was breaking up extensively. In much of Europe it was being replaced by a number of communities organised as independent units that functioned, both internally and externally, on the basis of monetary exchange. Feudal social life was being abandoned; the overlord and the serf were being replaced by the landlord and the rent-paying tenant. The merchant and the industrialist, who had never been closely integrated into feudal society, took their appropriate place in a new type of society that was organised with the broad aim of ensuring its own economic advancement and of providing a means by which individual persons could accumulate wealth.

The Italian cities had been pioneers in this process. Their citizens had created for themselves, by means of trade, a considerable degree of prosperity and leisure. They lived in a fairly cultured environment, and were remarkably free from feudal ties or papal influence. By the 13th century great interest had been aroused by the translation of works of Aristotle direct from Greek. In the 14th some of the people in these cities began to study Greek works concerned withnon-technical matters such as drama, history and the general Greek attitude to life (pp.105-109). They found, to their surprise, that the citizens of early Athens had assumed that a man should live life in all its aspects to the full. He should develop to the utmost his sensitivity, and his intellectual and physical capacities, thus attaining a well-developed body and personality. His reasoning should be used to this end, and his society organised in this light. This philosophy, which became known as humanism, differed greatly from Christian asceticism and its idea of original sin. It came to the Italians as a revelation.

This humanist viewpoint appealed to the successful men of commerce in the Italian cities. They felt that they themselves were self-made men who had, by their capacity and enterprise, attained their special status; they had themselves applied their human potentialities to some purpose. They used part of their funds to assemble libraries of ancient manuscripts and to collect ancient works of art. This patronage afforded opportunities for scholars and artists to study these manifestations of the older humanism and to express its contemporary significance in terms of the fresh art that it inspired. Here also prosperity and leisure were providing opportunities for living life to the full. The early cultured Greeks had ultimately owed their privileged position and their relative leisure to living in a very unequal society which included slavery; the equivalent Italians owed it to their special position within a major trading complex. Doubtless neither would have been inclined to view their position in this light.

The growth of humanism complicated the position of the Church. Today the term humanism is associated with agnosticism, but the Renaissance humanists believed that God had meant man (the male of the species) to employ to the full the faculties that He had bestowed on him, and accordingly they were impatient with the emphasis that the Church was placing on asceticism. This impatience showed itself particularly in northern Europe. Some humanists, being linguistic scholars, made a firsthand study of authoritative documents and drew their own conclusions from them. Erasmus, a Dutchman, made his own translation of the Bible, from Greek to Latin. Biblical criticism was beginning. Naturally the humanists, like so many others, were critical of some of the Church's current activities, such as selling pardons for sins (i.e. 'indulgencies'). Printing enabled their criticisms to be disseminated widely. The humanists did not wish to issue a direct challenge to the Church, but rather by reasoning to bring about its gradual reformation, and particularly to encourage it to accept and rejoice in the zest for life, with its urge to strive, to understand and to achieve, which is inherent in the human being.

The aim of Martin Luther, in Germany, was quite different. He was not a humanist. His primary interest was in religion. For him, the Church of Christ should be simple. The bureaucracy of the Church of Rome, its capacity to levy taxes, its wealth, its power, its pomp and above all its issue of indulgencies were wholly wrong. In 1517 he nailed his nintey-seven theses on indulgencies to the church door in Wurtemburg. In the past it had usually been easy for the Church to deal with heretics. But times had changed. Luther had the power of some of the German states on his side. The power of print was also on his side; his theses were widely disseminated by this means. He translated and printed a German language version of the Bible, so that anyone who was able to read could study it and reach his or her own conclusions. Lastly the wide prevailing discontent with the Roman Church was on Luther's side, a discontent which the humanists had themselves done much to rationalise and make articulate. Thus for various reasons the Church was unable to quell the movement that Luther had set in motion. The eventual result was that instead of a single Church there were two principal Churches in Europe, Protestant and Roman, which were for a long time bitterly opposed to one another. The humanists, who had stood for toleration and a tolerant form of Christianity, were ground between these two intolerancies.

The growing commercial drive had thus led, rather indirectly, to a major challenge to the contemporary Christian Church. This same drive also emphasised the importance of being able to resume trade directly with the East. As noted, improved methods of navigation now enabled ships to keep on course when out of sight of land. The recovery in 1406 of Ptolemy's 'Geographica', with its emphasis on latitude and longitude, emphasised the importance of the relation between exploration and the crucial problem of overall world geography. In 1437 Prince Henry of Portugal began to organise a systematic exploration of the west coast of Africa. He hoped to obtain tropical products from Africa itself, and he may also have hoped that his ships would be able to sail round the African land mass, the extent of which was unknown, and so reach India by that route. The increasing prosperity enabled him to set aside adequate resources. The explosive power of gunpowder, discovered in China, had become known to the West in the 13th century; here metal workers had been able to design and construct firearms by means of which this force could be used to propel a projectile along a pre-determined course. This achievement revolutionised warfare. It meant that the Portugese sailors, being alone equipped with guns, possessed an armament that could be used, or abused, in whatever contingencies they encountered.

Sixty years of cumulative exploration and consolidation gradually carried the Portugese further south until eventually they were able to round the southern tip of Africa. They then travelled some distance up its east coast before sailing across the Indian Ocean, so reaching India by that route. India was at that time a reasonably peaceful and prosperous region. In the south there was a Hindu population, and in the north a Moslem one. The maritime trade of the whole of the East had for many centuries been in the hands of the Moslems (p.121); they had established a chain of depots that included Alexandria and Aden, Ormuz in Iran, Calicut in the south of India, Malacca in Malaya, and Canton in China. However the Portugese did not make use of their independent access to the East, which they had gained so competently, in order to share in the trading complex that already existed there. Instead they employed their monopoly of artillery quite ruthlessly in order to destroy the Moslem trading complex and to begin the subjugation of the indigenous people of the area. Thus they established by force, for instance in the relevant parts of India, a situation in which they were able to take much and give little in return, in the course of which they greatly disrupted the lives of communities that were being subjected to enforced intrusion by very different people from a very different land.

Christopher Columbus undertook a different venture. If the Earth was a sphere, as early Greek studies had indicated, then if one sailed far enough westwards one should pass right round it, and so reach India from the other side. For this project he received some inadequate backing from the Spanish Court. In 1492 he sailed out into the unknown heart of the Atlantic Ocean. The prevailing wind was favourable, and after three weeks easy sailing he reached a group of islands which he believed to be in the Indian area, and which he accordingly called the West Indies. However it later became apparent that these islands were in no sense Indian; they lay off the coast of a vast and wholly unexpected continent which became known as America. Columbus had searched for a route to India, but instead he had found a veritable 'New World' which barred the way.

In 1519 Ferdinand Magellan left Spain to try to complete the task that Columbus had begun. He succeeded in sailing round the tip of South America, and thus reached and eventually crossed the Pacific Ocean. He was killed in a petty quarrel in the Philippines, but one of his ships completed the journey by rounding Africa and returning to Seville three years after it had left. For the first time the Earth had been circumnavigated. The spherical nature of the structure on which people dwelt had thus been demonstrated beyond all doubt by their ability to pass completely round it.

It is clear that by about 1500 the West had attained a dominant position in the world. It is less clear just how this came about. The fall of Rome had left society in the West fragmented. It had largely lost its communications, its administration, its towns and its trade. On the other hand China, India and the Byzantium portion of the Roman Empire had remained fairly prosperous, civilised and urbanised, as also had the areas that later became influenced by Moslem thought. In the West trade and city life gradually developed again, but this in itself would have done no more than bring this region to a condition that had already long existed elsewhere; it would not explain the subsequent Western relative ascendancy. Nor is it easily explained in terms of knowledge of basic processes. In this matter the Moslems had been reasonably well-informed at a time when the West were making little progress. Similarly paper making, block printing, gunpowder and the magnetic needle were known in China, and reached the West only secondarily from there. Where the West did differ was in its fragmentation and subsequent resynthesis. Elsewhere life had continued in much the same pattern since time immemorial; there had been little incentive to change. But in the West relationships and ways of doing things had had to be worked out afresh. Prosperity ultimately depended on productivity, and in the West this could be increased only by making better use of the limited manpower available. The emphasis was therefore on making more efficient use of the knowledge that was available. This could be achieved in various ways. Educated men might help by improving administrative procedures, by translating Arab texts, or by scanning news from China. Educated men of commerce could devise letters of credit and improve methods of bookkeeping, thereby creating financial machinery by means of which money could be used more effectively for productive purposes. Lastly it was manual workers, whether rural or industrial, who had direct experience of routine productive processes, and who would on occasion conceive ways in which these might be improved. It was in the interest of all concerned that educated men should use their wider knowledge and their influence to encourage and help local manual workers to transform any promising ideas into efficient new technical procedures. Here hand and brain were cooperating, even though the hands and the brains might sometimes belong to different sets of people. The practice of cooperation made people more disciplined, punctual and reliable. A growing body of information and ideas was freely communicated and gradually became part of the common Western heritage. The minds of the more effective individuals were aware of many things and of their interactions and implications. Such people were mentally alert and flexible, and they could understand one another and work together well. Western society had, for its time, become remarkably competent.

The achievements in Western societies seem therefore to have been due not so much to the discovery of things essentially new, but more to the tendency resolutely to pursue the possible practical applications of things already known in principle. The magnetic needle had long been known in China, but it was in the West that it came to be utilised in a compass for navigation. Gunpowder had also been known elsewhere, but in Western hands it led to the construction of efficient guns. Block printing was also known; its development in the West led to mass printing by means of moveable cast metal type. A feature of the situation in the West was the positive feedback element, with either mental or physical devolpments in one field encouraging further developments in others, so giving rise to a chain reaction. Comparable situations were probably rare elsewhere. They gave to the West its uniqueness and its relative ascendancy.

This relative position, already attained by about 1500, underwent no great change during the next two centuries. In the England that followed the 'Revolutionary Settlement' of 1688-89 the social climate became very suitable for enterprising activities of various kinds. The Settlement had brought a reasonable degree of freedom of speech, of the press and of religious practice, and thus in general an atmosphere of toleration. Monarchy and Parliament, and individuals engaged in agriculture, commerce and industry, and also in the field of science that was beginning to develop, were able to work together in broad harmony. Ideas that were generated by these activities were free to interact and by their interplay to demonstrate their degree of shortterm relevance. It was a period when energetic men with an enquiring turn of mind, and often with little formal education, could use their skills to develop in their workshops new methods of procedure that were relevant to commerce or technology, or in some cases to science.

However this situation, which appeared so tranquil, in fact led to fundamental changes during the second half of the 18th century. In the past the energy needed for productive activities had been obtained mainly by burning wood. The forests that had been providing the wood were of course in principle renewable, but were in fact being depleted much faster than they were replaced. Timber was therefore becoming scarce. This position was revolutionised by the discovery that the fossilised wood known as coal could be used instead of wood to fuel industrial processes. At first only horse-drawn vehicles were available to transport this coal from where it outcroped to where it was required, and the roads that these vehicles had to traverse became little more than quagmires during winter. However where the geography and geology were suitable this difficulty could be largely overcome by the construction of canals; these provided a satisfactory, but slow, means of transporting heavy bulky materials such as coal.

These developments became linked with other aspects of the contemporary scene. In the south of the United States the labour of slaves, originally forceably removed from their homelands in Africa, was being employed to grow large quantities of cheap cotton. Much of this was imported into Britain, where it was used to produce cloth that was much cheaper than that previously made from wool. In this colonial era the potential overseas market for this cheap cloth was immense; it could not possibly be met by the methods of spinning and weaving that were currently being practised as cottage industries in Britain. The profits that might result from successful innovations were therefore huge. Presently enterprising craftsmen were devising improved frames for spinning and weaving the cotton. Furthermore they found ways of using mechanical power derived from water mills to drive these devices.

This source of power was however soon superseded. Steam pumps were already being widely used to remove water that had accumulated in mines. After James Watt had introduced the innovation of cooling the steam in an independent condenser, and after the backward and forward motion produced by these pumps had been converted into rotary motion, then steam power derived from burning coal could be used to turn factory wheels. This had many advantages over power derived from falling water, not least that a factory did not need to be located beside a suitable stream. Thus by the early years of the 19th century textile factories powered by steam produced by burning coal had become a characteristic feature of the new British industrial scene. The introduction of various innovations of this kind made possible the mass production of cheap cotton goods, and this in turn enabled the immense potential markets that existed in India and elsewhere to be exploited profitably.

Two main categories of people were concerned with these new factory enterprises. On the one hand there were the owners whose capital had financed their construction. These were initially the same type of enterprising pioneers as those who had developed the new methods of spinning and weaving in the first place. It was to the advantage of these owners to pay as little as possible for the production of as many goods as possible, and to sell the latter for as much as possible. The initial generation of a head of steam was costly, so it was important that a factory should be kept running continuously. The employees were therefore required to work a shift system. The less they were paid, and the longer the hours they worked, the greater would be the owner's profit. Women and children, with their nimble fingers, were at least as capable of doing this work as men, and they could be paid less. An owner did not need to concern himself much about the safety or welfare of his employees. His commanding position arose because the whole situation was new, and no collective organization existed, or was at that time permitted, which could bargain about wages or conditions of work. Indeed the prevailing view was that the way in which factories were organised should be determined solely by the operation of the free market; any restrictions were undesirable. This view was rationalised by Adam Smith in his book 'The Wealth of Nations', which maintained that goods, which by definition included the labour that was hired, should be bought at the lowest possible price, and sold at the highest. This would produce the greatest possible profit which, ploughed back as capital, would enable the industry to grow as rapidly as possible. Development would thus be cumulative, and prosperity assured. The employee was thought to have no cause for complaint, for he did not have to accept a contract that was freely entered into between himself and the owner.

On the other hand the people who became the workers in this remarkable system were in a very different position. The former cottage industries could not possibly compete with the quantity of goods produced by these new factories, and were therefore no longer viable. The people who had been working in them therefore became workless, and had either to starve or to accept whatever conditions they were offered by the new factory system. They found themselves crowded into insanitary houses constructed in great density in the immediate vicinity of the factories. The crowding was inevitable since, in the absence of other means of transport, all 'hands' had to be housed within walking distance of the factory. The hours and conditions under which they worked seem today almost incredible; also their wages barely sufficed to provide them with the necessities of life. Malnutrition, infectious diseases, accidents and overwork all took their toll, and there was virtually no opportunity for relaxation or the cultivation of other interests. Their lifeexpectancy was much less than that of the welltodo.

The consequence that flowed from this 'Industrial Revolution' have played a considerable part in creating the social scene in which we live today. The workers, faced with much opposition, could not rapidly organise a movement that was able to press effectively for the major reforms that were so urgently required. However gradually, in the course of a century or more, they did procure increased wages, the legality of trade unions, safer and more satisfactory conditions at work, the right to vote, opportunities for education, and the introduction of medical and insurance services.

This new system of production also had secondary consequences. From the outset the factory towns required a continuous supply of cotton, fuel and food; also the finished products had to be removed. At first, as already noted, this transport was aided by the construction of canals. However during the first half of the 19th century improved engineering techniques enabled steam power to be used to turn the wheels that could drive a locomotive along a metal track. The resulting railway system greatly eased the transport problems, as regards coal, machinery, goods and persons. The railways were also highly profitable, and by about 1880 there was a network of them in most parts of Britain.

The construction of railways led in turn to further consequences. The production of engines, wagons, rails and associated equipment required the development of a major new industry. Large numbers of people in the West gained experience in the production and handling of steel in bulk, and in its conversion into precise and refined machinery. They were living in an environment in which they became trained in the art of devising and using lathes that could produce machine-parts that were virtually identical with one another, and therefore interchangeable. Mass production in factories was becoming a normal procedure, instead of being almost confined to textiles. All this was affecting the size, complexity and precision of the goods that were being produced, and the abilities and outlook of those people who became experienced in the making of these goods, and also the lives and attitudes of the people who had an opportunity of using them. In short, the nature and environment of whole communities was being in some degree transformed by this mechanisation of production. The consequences of the industrial revolution were extending.

Steam engines were used to propel ocean-going ships at about the same time as trains. These steam ships could be larger, could carry heavier loads and were more reliable than the sailing ships that they replaced. The overall result was that bulky goods, such as coal, wheat and mineral fertiliser, could be transported overland by rail, and then across the sea by ship, and so be moved rather easily to almost any part of the Earth. This meant, for example, that wheat grown on the newly-opened plains of North America could now be used to provide cheap food for the populations of industrial Europe, which were rapidly increasing. The various regions of the world were becoming less isolated, less self-sufficient and therefore more economically interdependent.

This industrial revolution, initiated in Britain but soon spreading throughout the West, naturally further increased its relative ascendancy over other regions of the world. And at the same time it was causing the different regions of the world to become more closely interrelated. About 1890 people therefore tended to assume that the West would continue to become increasingly prosperous, and also that this prosperity would gradually filter downwards from rich to poor, and similarly from Western to non-Western regions. The economic barometer on planet Earth thus seemed set fair a hundred years ago.

Chapter 14

SCIENCE BECOMES IMPORTANT

The development of science made an important contribution to the Western scene. Aristotle's broad synthesis (p.110) had become widely accepted by about 1300; it provided the West with a common conceptual framework concerning the overall nature of things. This work had covered the whole field of natural philosophy and presented its conclusions as a coherent whole. His intention was not to give people increased control over natural processes; to Aristotle that did not seem relevant, for he had been living in a society in which slaves and other menial workers could do all that was required. His aim had been to provide an account of the uniformity that was assumed to underlie the diversity that was actually observed. His system therefore had the advantage of providing a unified theory which brought to people's minds problems of interrelationships between different areas of human experience. It also gave a method of procedure, through observation and deduction. However it also had serious disadvantages. It did not look at how things happened, but why they happened in terms of underlying 'Forms'. It advocated a line of inquiry that tended to be sterile.

The broad validity of Aristotle's system was at first accepted in the West, but it was now received with fresh eyes and active minds. It was regarded as a system that could be elaborated, improved and built upon; it was not a weary and final answer to the problems of the Universe, as it had often been seen in late classical times. By about 1350 its assimilation was complete and a critical approach was beginning to develop. Technical activities provided one source of criticism, at least by implication. The smith who heated iron to a bloom and forged it with a hammer was performing an experiment. He was doing it for practical reasons, not to test a theory, but nevertheless he was, as it was then expressed, turning nature from her course, and was thus providing experimental evidence concerning the nature of things. If he used instead a bellows driven by water power and thus obtained a higher temperature at which the iron melted, he was performing a further experiment which threw fresh light on the behaviour of iron. Being a technician, he would have been dealing with a particular sample of iron that he wished to shape in a particular way; his experiences and innovations arose as a result of working at a particular task. He had no use for vague concepts of 'Forms'. His approach was directly contrary to that of Aristotle, and this was precisely because Aristotle had ignored the evidences afforded by the activities of working men. Yet it was due to people like smiths that technology had been making rapid progress in the West. Furthermore, this approach was clearly useful, for it led to technical improvements that gave rise to increasing prosperity.

Attention to the living world pointed in the same direction. The correct identification of plants was important for medicinal reasons, and plants could best be described by means of drawings, and to draw accurately required precise attention to the structure and nature of the particular plant in question. Surgery required similar precise knowledge of the structure of the human body. Human dissection began in the universities of Italy. The humanist movement that arose in that country, with its interest in human potentialities, in the form and grace of the human body, and its representation by means of art, also helped to foster the new anatomy. In 1543 Vesalius published his 'Da Fabrica', a precise and detailed account of the human body, described layer by layer, and organ by organ, and beautifully illustrated. Precise technical work brought progress.

There were also instances where Aristotle's system did not seem to be consistent with rational thought. He had asserted that a heavy body would fall faster than a lighter one. But it was pointed out, with typical scholastic logic, that if two canon balls were joined together with a rod they would surely fall at the same rate as the two balls in isolation, even though their joined structure was heavier.

Astronomy raised particularly difficult problems. Here Aristotle had broadly followed the concepts of Plato. He believed that the Earth formed the immoveable core of the centre of the Universe. The apparent movement round it of the stars, the Sun and planets was assumed to be circular, largely because Pythagoras, and Plato after him, had regarded the sphere as a perfect figure. In fact, however, this presumed uniform rate of circular motion around a stationary Earth did not explain the observed movements of the planets satisfactorily, and in particular it did not account for their occasional apparent brief backward movements (retrogradations) as compared with their previous courses. This was important, not least because the discrepancies upset the proceedings of astrologers. In Hellenistic times the astronomer Ptolemy had tried to meet these difficulties by introducing secondary circles (epicycles) whose centres were based on the circumferences of primary circles. This was able to preserve the uniform circular motion, but only at the cost of extreme complexity.

In view of these difficulties the 16th century Polish mathematician Nicholas Copernicus explored the possibility that an Earth which rotated daily on its axis and made a yearly journey round the Sun might yield a more satisfactory explanation. This brought no obvious improvement; this was because he, like others before him, was trying to explain in terms of uniform velocity and circular motion movements that were in fact neither uniform nor circular, as Johannes Kepler was soon to demonstrate. However before his death in 1543 Copernicus had shown that his revolutionary hypothesis gave results that were as satisfactory as the orthodox ones, and that it had the advantage of being simpler and more harmonious.

The old difficulties still remained. A generation later the astronomer Tycho Brahe was impressed by the need for observations more precise than any that had yet been made. The King of Denmark built an observatory for him. His instruments were of the same general type as those used previously, but much larger and more finely graduated. He was careful to define the limits of accuracy of his observations. This was a new and important departure; it meant that, within these limits, the accuracy of his measurements could be relied upon. He was careful also to make regular and systematic observations, which was another innovation, for previously measurements had usually been confined to occasions when the planets in question were particularly easy to observe. Brahe thus for the first time provided data that enabled the position of a planet to be traced along the whole course of its orbit.

The practical work of Tycho Brahe was turned to theoretical account by Johannes Kepler. He was an astrologer by profession and Pythagorean in outlook, and he sought passionately to understand the mathematical harmonies that he sensed to be incorporated in the matrix of the Universe. He went to Brahe, who alone could provide the wealth of accurate data that were needed, and when Brahe died in 1601 he took charge of his material. He found, to his surprise, that the orbit of Mars could not be regarded as circular; the divergence was not great, but it was greater than the known degree of inaccuracy to which Brahe's measurements were subject, and it had therefore to be regarded as significant. Eventually, in 1609, after immense labour, he succeeded in showing that the observed courses of the planets could be accounted for if they moved round the Sun in ellipses, not circles, and if their speeds increased when they were nearer to the Sun; this therefore also meant that their rates of motion were not uniform. The uniform circular motion, assumed by Plato and

Aristotle and thereafter through the ages, therefore did not accord with careful calculations based on precise and systematic observations. Kepler conceived instead a different pattern of motion which did correspond with the observations; this motion was neither uniform nor circular, but it was no less harmonious.

However, in spite of such discrepancies, the Aristotelian framework of ideas was accepted almost universally until about 1600. One reason was that there was no comparable system that could take its place. Another was that the presumed central position of the Earth had become incorporated into basic Christian doctrine. A further difficulty was that the system was so closely integrated that the destruction of any major supporting pillar would cause the collapse of the whole edifice; consequently any investigator whose experience in a particular field was leading towards this result was inclined to recoil in alarm, believing that it must surely be himself, not Aristotle, who was erroneous.

It was the discoveries and the perseverance of Galileo Galilei that broke through this barrier. He had been trained in mathematics, and then taught in the trading cities of north Italy, where he came to think that the work of natural philosophers should have practical applications. He took Archimedes (p.113) with his experimental approach to natural phenomena as his model. He seems to have been proficient enough as a craftsman to construct his own experimental apparatus. This single person personified many of the prevailing trends, both mental and manual, that were by implication critical of Aristotle's system.

Galileo concerned himself mainly with practical problems until about 1602, when he was 38 years old. At this time his interest in pendulums seems to have led him to wonder about the way in which objects fall. He ignored all questions about the nature of gravitation or the reason for its existence, confining his attention to the effects that it produced. The time it took for objects to fall vertically was too short for satisfactory measurement. He slowed the process down by, instead, rolling balls down gently inclined planes. This enabled him to measure the rate of fall reasonably accurately with the simple means at his disposal.

In this work Galileo was primarily interested in interrelationships, and he was able to show by his experiments that, as he had already surmised conceptually from general principles, the distance that a ball travelled down the inclined plane was approximately equal to the square of the length of time that it had been in motion. He appreciated that in any case complete conformity with his conceptual calculations could not be expected; for one thing his measurements were necessarily not very precise, and in addition the balls were not moving altogether freely, being subject to various complicating factors such as air resistance and the resistance between them and the inclined plane.

By this procedure Galileo first isolated one particular problem, namely that of falling bodies, and had then concentrated his attention on one particular aspect of this problem, namely the relation between time taken and distance travelled. He did this because he had perceived that it might be crucial to an understanding of the process of falling. In addition he had conceived within his mind the possible nature of this relationship, and his experiment had demonstrated that his surmise was indeed correct. It had also demonstrated that this relationship could be expressed in simple mathematical terms. He appreciated that when these studies were applied to a particular problem, such as for example calculating the course that would be followed by a projectile propelled from a gun, then it was necessary to take into account not only the basic relationship that had already been established, but also to estimate the effect of air resistance, wind pressure and other minor factors that had so far been intentionally ignored. The crucial difference between the approach of Galileo and that of Aristotle was therefore that the former demonstrated experimentally that his concepts did conform to the actual motion of falling bodies, and that they could therefore be relied upon.

Galileo made further use of this simple experiment. When the balls had reached the bottom of the inclined plane they were allowed to continue on their course across a horizontal surface. Here the force of gravitation was no longer causing them to accelerate, and they were gradually brought to a halt by air resistance and friction. Galileo found that their capacity to resist such causes—that is to say, their momentum or force of motion—could again be expressed by a simple mathematical relationship. It was equivalent to their weight (or, more strictly, their mass) multiplied by their rate of motion (i.e. their velocity). This again was an abstraction, and again it represented the essence of the matter. Thus by interrelating thought and experiment Galileo came to realise that it was not motion itself, but a change in motion, that required the action of a force. This was the reverse of Aristotle's supposition, and it showed his whole system of mechanics to be invalid.

It was in these general circumstances that Galileo heard, in July 1609, that lens-grinders in Holland had discovered a means by which, if light from a distant object was passed through a pair of lenses that had been shaped in an appropriate manner, the object in question was made to appear as though close at hand. By December of that year he had himself prepared a comparable device.

The results were dramatic. He could clearly see that the surface of the Moon was covered by mountains and plains, and that the former cast shadows on the latter. This was very disturbing to orthodox opinion, for Aristotle had maintained that the Moon and other heavenly bodies were absolutely spherical and were made of a substance superior and wholly different to that of the Earth. Galileo's telescope now showed clearly that that was not the case; in principle the Moon resembled the Earth. The telescope also showed that the vague glow of the Milky Way was due to thousands of separate stars. In January 1610 he saw that the planet Jupiter had four satellites revolving round it; this demonstrated that the Earth was certainly not the centre of all things, since for these satellites it was Jupiter that was the central body. In September of the same year he saw that the planet Venus had phases similar to those of the Moon. This showed beyond doubt that Venus revolved round the Sun, not round the Earth; in this respect, therefore, it was Copernicus who had been correct, and Aristotle who had erred.

Galileo had therefore shown that Aristotle's system of mechanics was wholly incorrect at numerous crucial points, and the invention of the telescope had demonstrated that his interpretation of the heavens was equally defective. Galileo had thus removed not merely one major pillar, but several, from the Aristotelian edifice, and he considered that the whole structure was unsound and needed to be demolished. He knew that his own method of interweaving theory and practice led to positive and reliable results. Thus he felt that his approach to natural phenomena was far superior to that of Aristotle, not so much for what it had already accomplished as for the new horizons that it had brought to view. He could offer no broad integrated system to substitute for that of Aristotle, but he had shown that a growing system of natural philosophy, this time firmly based on experimental evidence, could gradually be constructed. In his later years he therefore embarked on a full-scale attack on the whole body of Aristotelian thought. He did not just state a contrary case, but struggled actively, by wit and ridicule as well as reason, to win people's minds to the new ideas that he had himself attained slowly and with so much effort. As is well known, this campaign provoked an extreme reaction from the Catholic Church. However Galileo was also beginning to sense that the relationships he was uncovering were themselves interrelated, being parts of a larger integrated framework. Humankind was obtaining a first glimpse of the underlying workings of the Universe. One could, he thought, learn about God, the Creator of the Universe, as well through the book of Nature as through the Bible.

The importance of Galileo's work lay not so much in what he had discovered as in the method of procedure that he had introduced. He had shown that one could isolate a particular matter for attention, that by an act of imagination one might form an hypothesis regarding the essence of the problem that it presented, and that this concept could then be tested experimentally to see whether its actual behaviour did in fact correspond with that anticipated by the theory. If it did not, then this theory would need to be abandoned. But if it did, then the concept could be tentatively regarded as correct, and some further consequences that should flow from it could in their turn be tested. Publication could give all concerned access to the results obtained, so enabling others to form their own ideas, and conduct their own experiments, and so reach, and publish, their own conclusions. This new method of procedure, known as scientific, could advance step by step; at each step the relevance of the mental concepts would need to be checked by experiments, and only those that were found to conform to the behaviour of the actual world would be retained. Theory and practice would thus be continually interacting, and imagination would, for these purposes, remain closely linked with actuality. The human understanding of the nature of the world could thus, by means of science, progress in a systematic manner. Galileo himself had naturally not been able to travel far along this road, but he had pioneered a way that others could freely follow.

The privilege of forging ahead in Galileo's particular field fell principally to Isaac Newton. It is said that in his youth an apple falling from a tree had stimulated him to wonder how far the force of gravitation extended outwards into space. Could it be this same force that was continually drawing the Moon into a curved orbit round the Earth and so preventing it from flying off at a tangent? Kepler had shown that the planets moved in ellipses round the Sun. Newton was able to show that if a planet was subject to an attraction which decreased as the square of the distance between it and the Sun, then it would in fact traverse an elliptical course. He also succeeded in showing, though with difficulty and only by using the mathematical technique known as the infinitesimal calculus which he himself had been partially instrumental in creating, that the attractive force exerted by a massive sphere could be treated as though wholly concentrated at its centre. This was important, for it meant that he could now bring precision to the problem of Earth/Moon relationships. The distance between the centre of the Moon and the centre of the Earth was known to be about 60 times greater than that between the surface of the Earth and its centre. Hence, assuming that the inverse square law holds, the force acting on objects here on the Earth's surface should be 60^2, i.e. 3600, times greater than that acting on the Moon. In each second the Moon falls 0.0044 feet towards the Earth; it is this fall that draws it from a straight tangential course into a curved one that keeps it circling round the Earth. Objects on the surface of the Earth should therefore fall towards its centre at 3600 x 0.0044, i.e. 16 feet in a second. And this was known to be the rate at which things here on the Earth's surface do in fact fall. It followed therefore that it was the same cause, of unknown nature, called gravitation, that was responsible both for making apples fall to the ground and for keeping the Moon circling round the Earth.

Newton published the results of his work on this and similar matters in 1687. In this classic publication he did far more than demonstrate the underlying mechanics of the Solar System, for he was also able to deduce the existence of a universal principle, namely that every particle of matter behaves as though it attracts every other particle with a force that is directly proportional to the product of their masses and inversely proportional to the square of the distance between them. Subsequent observations and experiments have always accorded with this principle, apart from certain minute discrepancies which can be accounted for in terms of Einstein's even more wide-ranging 'Theory of General Relativity'. Thus Newton, using more intensively the kind of procedure that Galileo had invented, had pushed his way through a maze of complexity and discovered a principle that was apparently universal in its scope. Again it could be defined with mathematical exactitude, and again, though essentially simple, it was not obvious to common sense. Newton had achieved the first great synthesis in the history of science, for he had demonstrated that a whole aspect of the Universe behaved in accordance with a simple underlying principle. The method of procedure developed by Galileo was proving wonderfully successful.

Newton had applied the methods of Galileo to one particular field. However different features of the world each have their own natures, and present their special problems. Newton's work, for instance, was of no particular help in trying to understand the nature of the materials that comprise the Earth and are present everywhere around us. Experiments in this field over several centuries, first by the Moslems and later also in the West, were based on little useful insight. While empirical knowledge had increased, the nature of the underlying problems still remained obscure. It is interesting that just as Tycho Brahe's careful measurements had led to the concepts that eventually revealed the nature of the Solar System, so it was also careful measurements, in this case of the relative weights of materials before and after they had reacted chemically, that led to the crucial breakthrough. This comparison of relevant weights was the work of Antoine Lavoisier, and its possible implications were grasped by a school teacher, John Dalton, in 1803; this was more than a hundred years after Isaac Newton had published his major synthesis. Dalton was applying in a modified form the early Greek suggestion that elements consist of minute particles or atoms, and that all the atoms of any one element are changeless, indivisible, and identical in size and weight, but are different from atoms of any other element. He realised that, on this assumption, the precise relative weights that Lavoisier had found to be involved in chemical changes could be explained if small whole numbers of atoms of one element had combined with similar whole numbers of those of another element, so forming molecules.

This suggestion did work. It enabled chemists to interpret changes brought about by experiments or by industrial processes in terms of the way in which the postulated underlying atoms were linked with one another. On this basis they became able to explain observations and experiments, and to make and test predictions, with insight based on an understanding of the underlying principles. These practical activities in turn helped the conceptual framework to expand. Chemistry had become able to develop into a science. It soon became possible to calculate the relative weights of the atoms of the different kinds of elements. It was noticed that when the elements were arranged in the order of their ascending atomic weights, elements with similar characteristics appeared at regular intervals along the series. This 'Periodic Table' suggested that there were underlying regularities, of a nature as yet unknown, in the structure of the different kinds of atoms. An expanding network of interrelationships was emerging, along with indications of the existence of still deeper and more fundamental ones whose nature still remained unknown.

Light presented its own quite different problems. It exists all around us, and it is the medium that enables us to see. Yet it is intangible. At an early stage Newton had shown that if a narrow beam of sunlight is passed through a triangular glass prism it emerges as a series of bands of coloured light. This series, which is violet at one end and red at the other, is known as a spectrum. If the colours in question are painted on a cardboard disc and if this is rotated rapidly, the individual colours can no longer be seen and the disc appears to be white. These observations suggested that the white light that is radiated from the Sun consists of a number of different colours which are indistinguishably

blended, and that when this light is passed through a prism the various coloured components are in some way separated from one another.

In this case it was the nature of waves that provided the crucial clue. If
waves are set in motion at two points in a pool of water they will spread outwards until they meet. In the zone of contact there are positions where the
crest of one wave meets the trough of another, so that they cancel one another
and a calm waveless area results. There are positions nearby where two crests
meet and augment one another. The result is that positions with larger waves
alternate with those where there are none. In 1801 Thomas Young redis-
covered that if an appropriate source of light was passed through two separate
minute holes in a way that caused the light from these two sources to overlap
then a similar alternation, resulting in this case in dark and bright bands, was
produced. The dark lines, called interference bands, were positions where, as
in the pool, the two systems of waves had interfered with and cancelled one
another. This indicated that light behaves as a form of wave motion. It became
possible to calculate the intervals separating the crests of successive waves, this
being known as the wavelength of the light. The spectrum of sunlight was
found to grade from a wavelength of 0.00004 cm. at its violet end to 0.00008
cm. at its red end. Clearly, therefore, the colours of the spectrum were associated with light of different wavelengths. If the white light radiated by the Sun
was passed through a prism these wavelengths became separated because the
shortest were bent (i.e. refracted) by the glass the most, and the longest the
least; the light therefore emerged as a regular series of coloured bands that are
associated with a steady gradation in wavelength.

As regards living organisms, most of the processes that make life possible
are of a chemical nature, so satisfactory progress could not be made until
chemistry, and particularly the chemistry of carbon compounds, had been
placed on a firm foundation. This did not take place until comparatively late,
so it was not until the 1860s that it became possible to begin to appreciate the
way in which the bodies of plants and animals functioned, and so to establish a
science of biology.

There was also the problem of understanding the nature of rock formations.
It happened that the construction of the canals that were so urgently needed
for transport during the Industrial Revolution (p.131) required careful attention to the geological structures that were encountered. William Smith, who
was one of the engineers, provided descriptions and illustrations of the types
of fossils that were characteristic of the different rock formations. This information helped Charles Lyell to develop his 'Principles of Geology' during the

early 1830s. In this work he showed that the strata of the familiar sedimentary rocks had been formed from the remains of accumulated layers of silt deposited at the bottom of earlier seas. The fossils were the remains of living organisms that had, after their death, fallen to the bottom and become embedded in this silt. These strata succeeded one another in sequence, the oldest at the bottom. The succession of these sedimentary rocks therefore demonstrated the relative ages of the sediments, and therefore also those of the plants and animals whose fossilised remains had become entombed there. The evidence suggested that events during the past had proceeded in much the same fairly uniform manner as they do at present; clearly the periods of time involved were very long. Lyell's work thus provided the crucial conceptual framework on the basis of which a science of geology was able to develop.

Similarly there was the question of how the different kinds of plants and animals had come into existence. Charles Darwin had been considering matters of this kind when in the 1830s he had worked as Naturalist on 'HMS Beagle'. After returning home he happened to read 'An Essay on the Principle of Population' by Thomas Malthus, and it was apparently this that gave him the crucial insight that he needed. During the next twenty years he gathered a mass of information in support of his hypothesis that existing species of plants and animals had evolved from earlier forms as a result of natural selection. The book that resulted, entitled "The Origin of Species by means of Natural Selection" and published in 1859, provided a seminal concept that has illuminated all subsequent biological studies. In addition these concepts of Darwin and Lyell have conjointly enabled us gradually to understand the interrelated living and non-living processes that have together been shaping the nature of things on the surface of the Earth during the last 600 million years.

Science, in the form of a number of different disciplines, had therefore been created. The method had been to concentrate attention on particular aspects that are small enough to be manageable, to test the validity of relevant ideas about them, and by continuing along these lines to endeavour to reach conclusions about their underlying nature.

The Universe as thus revealed late in the nineteenth century seemed a very mechanical organization. There was Newton's clock-like Solar System. There was Dalton with his changeless, invisible and indivisible atoms that lay at the heart of all chemical changes. Animals were products of these changes, and so could be regarded as essentially machines. How then were we humans related to this situation? We knew ourselves not to be mere robots, and so we were regarded as wholly different from animals, perhaps because we each possessed a soul. In any case we were set apart, for it seemed obvious that it was humans who did the investigation and the interpretation, whereas it was nature that was investigated. The more we came to understand the mechanisms that underlay this very mechanistic world the more power we would have over it. Progress was inevitable. The world was our oyster. As regards science, as well as economics (p.134), the barometer seemed set fair a hundred years ago.

Chapter 15

THE LAST HUNDRED YEARS

The driving force behind the ascendancy of the West had been the encouragement of technical developments by a profit-oriented society. It had long been expected that science would assist technology, and from an early stage men of commerce had been inclined to give moral, and sometimes also financial, support to men of science, for they hoped that these investigations would eventually provide a means of improving the techniques on which they themselves depended. In fact, however, this scarcely occurred until the second half of the 19th century. By then some ideas derived from science were being incorporated into the procedures employed in engineering, agriculture and medicine. Also, more importantly, developments in science sometimes led to the creation of completely new types of industry. Synthetic dyes provided an early instance. A brightly-coloured organic compound that could be used as a dye had been produced by William Perkin in Britain in 1856. Chemists in Germany soon set about synthesising compounds of the same general type but in which the structure of the molecules had been modified in various minor ways. The result was a whole range of usable dyes of very different colours. Here, then, scientific imagination, acting in conjunction with some appreciation of the relevance of the molecules concerned, was being used by chemists to produce useful substances that had never previously existed on the Earth. A new type of industry grew out of this, which was based on scientific understanding of infrastructure as well as on technology. More recently a similar approach has been applied to other quite different types of molecules, which similarly led to the science-based industrial production of whole new ranges of, for example, plastics, pharmaceuticals and pesticides.

These particular science-based industries were an outcome of the synthesis of new types of molecules. There are numerous different ways in which science can be applied. Two other examples will be noted. First, science had shown that electric currents are produced by streams of electrons that have been caused to move between two points, usually along a wire. Around 1880 Thomas Edison succeeded in using an electric current to produce light in an incandescent lamp. The benefit derived from this 'electric light' was so great that soon dynamos, first constructed and used experimentally by Faraday some fifty years earlier, were being improved, manufactured and used to transform power derived from falling water, or from steam, into electricity. Electricity was then distributed from these power stations to, eventually, almost every home in Britain. The arrival of this electricity not only provided a source of light; it also meant that small electric motors could be used for running new kinds of household appliances, such as vacuum cleaners and refrigerators. Similar, but larger, motors could be used to power a workshop or a light industry. Machines of this type are small, compact and clean, and can be turned on or off at any moment as required. Small industries can be located in any place, urban or rural; they do not have to he close to a source of fuel. These new horizons were made possible by the creation of a new science-based electrical industry.

The history of the second example also began about the same time, when Alexander Bell succeeded in transmitting a human voice along a wire, so creating the first telephone. Irregular sound waves produced in a human larynx were converted into equivalent vibrations in a flat disc, which were then superimposed on an electric current which traversed a length of wire. At the receiving end a reverse series of changes took place. It was remarkable that such a complex pattern should have been preserved through such a complicated series of transformations. It suggested great potential versatility. This principle was used again when streams of electrons were induced to pass across an evacuated space, as in a thermionic valve, or through solids of the type known as semi-conductors. Such electrons move exceedingly fast; they have virtually no inertia; they can be diverted by an electromagnetic field, and respond immediately and precisely to any variations in that field. Since variations in most physical processes can be transformed into variations of this type, this is a highly versatile tool. An alliance of science and technology, making use of, for example, an ability to amplify weak currents and to transform patterns of light impulses into corresponding patterns of electron movement, has led to the creation of such devices as television, radar, computers, robots and electron microscopes, guidance systems for artificial satellites and for missiles and anti-missiles. The environment and general way of life of people, especially in the more developed regions of the world, has been changing fast as a result of this science-based electronic revolution.

Comparable changes were taking place in the biological sciences. Already in the 1850s Louis Pasteur was demonstrating that micro-organisms that cause fermentation of wine were not generated spontaneously from non-living matter, but by the multiplication of pre-existing organisms that were already present. He then went on to show that some important diseases were also caused by infections due to micro-organisms. This discovery provided the inspiration that led to Lord Lister's introduction of antiseptic measures for surgery; this, used in conjunction with anesthetics that had recently become available, converted surgery into a far less unsatisfactory procedure. Pasteur, Robert Koch and others were soon isolating and studying the particular types of micro-organisms that were responsible for particular diseases; they were then able to ascertain the nature of the processes involved in these diseases, and hence to recommend measures for preventing or treating them. These achievements also encouraged a concept that was in any case gaining ground, namely the importance of sanitation and cleanliness, and particularly of trying to ensure that drinking water did not become contaminated by sewage. Public health and medicine ware thus beginning to become science-based, based on some understanding of the underlying processes involved. Since then there have been innumerable further developments; among others vaccines, immunization, antibiotics, organ transplants and progress in the understanding of genetics.

In our most recent history, we have come to realise several important aspects in the development of our scientific knowledge. First, to obtain practical results from innovative concepts of this kind usually required cooperation between scientists and technologists. Science and technology have thus come to work together to bring very many projects to fruition; such conjoint activities are now called 'Research and Development', or, more briefly, R&D. Secondly, these activities have been taking place within the context of separate competing sovereign states, and these are based on a profit-oriented form of economic organization. Thirdly these states — along with their policies, their economic activities and their R&D — are organised by us human primates, each of us with our cerebral cortex and highly complex neural networks, and each equipped with our capacity for thought, speech and writing, and each with our individual experiences, memories and feelings. Fourthly, the present existence of seemingly almost unlimited reserves of fossil fuel, when used in conjunction with the very effective R&D, has given to us humans enormous potential power. We can, for instance, cover a whole country with a network of motorways, and can manufacture, maintain and fuel a multitude of vehicles that traverse them. Similarly we can create and employ machines that are easily able to push over, and uproot, every tree in a forest. Furthermore the prevailing economic system is such that personal profit, rather than the common good, can easily be a major factor in determining how in fact this power is used. And fifthly, we live on a planet that is quite small and has only limited resources and delicately poised dynamics. It is becoming essential to moderate our use of the excessive powers that we currently possess and learn to employ them in ways appropriate to the nature of this planet. The Earth is the only home that we humans have, and to the extent that we damage it we damage also our own future. We urgently need to acquire and widely distribute reliable information and understanding. Responsible decision-making, judgments and self-discipline are required. Our learning has not simplified the situation of humankind on planet Earth.

The various aspects of the planetary scene are closely interrelated and now form a complex network. Four of these are briefly noted below, namely (a) the socio-economic situation, (b) the environmental situation, (c) the human population situation and (d) the science situation.

(a) The socio-economic situation

During the Middle Ages the precursor of fully-developed capitalism had given those concerned the incentive to think, explore, innovate and act in ways directed towards their own advantage. By the 18th century those concerned were able to develop types of machinery and attitudes of mind that enabled them to apply the prevailing economic system very effectively to the mass production of goods. This, together with the discovery of the usefulness of fossil fuel, enabled the industrial revolution to progress rapidly during the 19th century. Later this, along with the growth of science and the associated R&D, encouraged Western states to overrun and assign to themselves great blocks of land in relatively undeveloped portions of the world, and to administer these colonies and use them as markets for the sale of the great flood of goods that they had now become able to produce.

This capitalist system, developing under its own momentum, has changed greatly during the present century. The more successful enterprises have become larger and more powerful, and have in many cases spread beyond the countries where they originated. These have become huge multinational organizations, often with interlocking branches in a number of different sovereign states. Subsequent R&D procedures have helped such organizations to deal with the many problems arising from their own increasing size and complexity. Competition between these giants and small companies in the same line of business may become reduced to zero as the latter get squeezed out of existence. This does not necessarily reduce the total number of companies, for new small ones are liable to become established in what they hope will prove favourable niches within an overall developing situation. Often some of the multinationals get taken over by others, and this results in the presence of even fewer, larger and more monopolistic giant enterprises.

The capitalist West has now been in operation long enough for its main strengths and weaknesses to be visible. Its principal strength is that it provides a means of producing and distributing goods efficiently, in quantity and cheaply. Among its numerous weaknesses is that its productive procedures are divisive and unfair. On the one hand there are the owners, who now included the various types of shareholders; decision-making rests with them, and surplus profits go to them. On the other hand there are the employees; they are excluded from decision-making and they receive merely wages. These distinctions do not depend on the ability, enthusiasm or hard work of the individuals concerned. Secondly, the distributive procedures certainly get the goods from factories and farms to shops, but once there its further distribution depends on how much money an individual has, and this may depend on factors that are outside his or her control. Thirdly, the system appears to he unable to avoid alternating periods of growth and recession. During the latter the jobs of millions of employees, including some of those employed to manage, are threatened. Fourthly, as noted above, there has been a tendency for industrial enterprises to become concentrated into relatively small numbers of very large units; this has largely eliminated the free and equal competition that used to be regarded as a major virtue of the system.

A further important feature is that the various sovereign states each have their separate economies, as do the individual organizations and individual persons within these states. Each state therefore competes with other states with a view to promoting its own particular financial interests. This competition dates back to the Sumerian communities, and thus to the very beginning of civilization (p.97). From that time onwards it has been a major factor leading to the creation of armed forces in such states, and thus to all the expense, wasted resources, and largely unjustified propaganda, and also the periodic outbreak of wars and the death, misery and disruption that these cause. It was competition between Germany and Britain for the control of potential markets in Africa and elsewhere, that, backed by their respective armed forces, led to the World War of 1914-1919. The destruction of millions of young human lives dispelled the illusion, prevalent at the end of the 19th century, that the

20th century would be plain sailing. The despair involved in this war also resulted in Russia's feudal and weakly capitalist society being displaced by a soviet socialist one in 1917. It also thus lead to this new type of state being regarded by the capitalist states as the great enemy, surpassing their enmities to one another. It was this kind of situation that induced the capitalist West, after the 1939-1945 war with its complex alignments, to create a large and varied stock of nuclear weapons, with the soviets duly following as soon as they were able, and then with each side stimulating the other to make ever greater efforts.

The world in the present century is characterised, as a result, by three principal types of society. These are the capitalist West, the Soviet sector and the 'Third World'. The first has already received some comment; we will know look at the second and third.

In the early days there were considerable hopes that the new community-conscious Soviet regions would have a bright future. However, particularly between 1945 and 1985, the USSR, which was its crucial component, underwent a disastrous deterioration. Probable reasons include, firstly, the exceedingly heavy human and material losses suffered during the 1939-45 war, and the Soviet unwillingness to accept from the USA anything equivalent to the Marshall Plan that could have contributed towards a rapid recovery. Secondly, competition with the capitalist world was fierce, and they used a large proportion of their very limited resources to create an armed force, both conventional and nuclear, that they hoped would be strong enough to discourage military intervention against them. Thirdly, in the absence of a profit motive they found it difficult to provide adequate rewards for enthusiasm and hard work. In the earlier phase the thought that one was helping to build a better society, and therefore a better life for all concerned, often sufficed. Later, however, cynicism more often prevailed, especially in farms and industrial enterprises that were too large and poorly organised to deal effectively with the individual human problems that arose. Lastly, and perhaps most crucial, they failed to develop any satisfactory way of making those in positions of power accountable; such people were therefore able to provide themselves with many privileges. Similarly there was no means of checking the spread of corruption and nepotism at these levels.

This phase was brought to an end in 1985 when Michael Gorbachov, on becoming General Secretary of the Communist Party, set about trying to introduce openness, democracy and essential reorganization into the administrative and productive apparatus. This bold move was clearly necessary, but it has inevitably led to much confusion and dislocation. The outcome, in 1997 with Boris Yeltsin at the helm, remains very uncertain.

One satisfactory aspect of these very recent changes is that the ordinary citizens in this vast former soviet area now have far more opportunity to think things out for themselves without interference, and to make use of a far wider range of information. There is however nothing in their recent history that can help to guide them on how to make wise and effective use of these new opportunities. Although the forms of social organization constructed during the earlier stages of their soviet history were highly community-conscious, this approach has had to be abandoned in favour of an unorganized scramble that involves its replacement by market forces. People may thus find themselves participating in the construction of what at present seems to be becoming merely another, and particularly inefficient, example of a capitalist economy. This could be unfortunate for two main reasons. Firstly, it may be that economies built on these Western market principles are bending, though more slowly and for different reasons, towards much the sane kind of chaos and disintegration as has overwhelmed the Soviet Union; in this event the people there will have jumped out of one frying-pan into another. Secondly, trying to view the matter in a world perspective, it may turn out that the best solution to our planetary situation is to develop economies that are constructed on the basis of a judicious blending of cooperation and competition. In this event a reasonably objective history of the early Soviet Union, at the tine when its development was still based largely on concepts of cooperation and public welfare, might at a later stage prove very helpful. In the past the writing of such a history has been almost impossible. In the near future it may again become impossible, if only because death and time will have further obscured the evidence. But in the present highly fluid phase it might perhaps be possible. As regards the overall picture, it seems that virtually the whole world, perhaps with the important exception of China, is becoming encompassed either in the capitalist world or in its Third World consequences. Thus the whole area north of the Rio Grande, of the Mediterranean and the Himalayas can be regarded as broadly capitalist-oriented, whereas most of the countries south of that line, apart from major capitalist outposts in Australia, New Zealand and South Africa, fall broadly within the Third World category. If both these major regions should collapse in chaos, then no alternative social systems, apart from China and the memory of the early Soviet experience, will be readily at hand.

The extensive decolonization of former colonial areas took place between about 1940 and 1960. The freedom granted was however principally political, with the result that many of the economic ties with their former imperial masters were still retained. This has left the peoples of these Third World countries, who comprise a large proportion of the human population on the Earth, in an unenviable position. The peasants, labourers and others in these countries have been learning, particularly through the medium of television, of the relatively luxurious standard of living enjoyed by typical people in the Western world; they have therefore come to expect that they should be able soon to attain a comparable standard. The local chiefs or others who then came to be in charge of their affairs usually had no previous experience of the techniques required for the successful mass production of goods; also they had neither the necessary machinery nor the capital that were needed to invest in its construction. It seems to have been widely assumed, when the former colonies became independent, that the development of post-colonial and other underdeveloped countries would follow much the same course as the countries in the West had done earlier. Western banks, and in particular the World Bank, were ready and eager to provide loans at interest on conditions that were based on this assumption.

In fact, however, the overall situation in these former colonies was very different. In the West industrialization had been a gradual process that had been taking place among populations in which many of the people had long been accustomed to dealing with materials in quite complex ways (p.130). Most people in the former colonial areas did not have this helpful type of prior experience. These states needed to undergo the process quickly, impatient as they were to participate in greater prosperity, and the need to be able to pay interest on the loans that had been granted to finance the proposed new projects and, in many cases, to deal with the problems arising from the rapid increase in the size of their populations. In the West the process had been more spontaneous and direct, for the ordinary people had themselves participated in it, and had to some extent initiated the new developments. All this had made assimilation of the new experiences relatively easy as, with plenty of time available, they could smoothly and gradually translate their own direct experiences into equivalent new patterns of activity within the neural networks in the cerebral cortices of the individuals concerned (p.130). They were able to spontaneously develop the changes in their overall thought processes and personalities that were appropriate to the new situations in which they were becoming involved. The Third World has had none of these advantages. The West has had the important additional advantage of having been first in the field. It had had an open market, indeed in the colonial areas a largely captive one, in which to sell the cheap goods that it was producing. The underdeveloped countries have had no such advantage, for in this competitive situation all the more obvious opportunities were already being exploited. The playing field on which the two sets of players operated have by no means been level.

There were a few non-Western countries, notably India, that were heirs to ancient civilizations that already had considerable productive resources and potentialities. A straightforward process of industrial development, sided by Western loans, was more easily achieved. In less developed countries Western banks' advisors preferred that loans to be used initially for developing and selling mineral resources, such as coal, iron, copper, gold, diamonds or uranium, where such raw materials were present in sufficient quantities. Third World countries lacking such resources were offered loans that would enable them to use their most fertile land for the intensive cultivation of crops, such as coffee and bananas, that required a tropical climate and that, being eagerly sought by Western markets, could readily be sold for cash. The production of these cash crops would, they thought, help the countries concerned to pay at least the interest on the loans, and at the same time would increase the technical and material consciousness of the people in the country concerned.

This last type of programme had some particularly unsatisfactory consequences. It inevitably meant that the country's most fertile land, being now used for cash crops, was no longer available for growing the crops that had previously fed the local people and also their compatriots in adjacent towns. Some of the farmers found employment in the new cash-crop industries, but many were forced by the new circumstances to leave their land and move to the shanty settlements that were coming into existence at the margins of the towns. These countries thus became, among other things, less self-supporting.

The situation in the rural areas became aggravated when agricultural research workers in the West began to breed strains of local food-plants, such as rice, that were far more productive than those previously used. The 'Green Revolution' brought trouble, as these new strains would not flourish unless provided with larger quantities of fertilizers and pesticides than the smaller peasant farmers could afford. The relatively large-scale farmers, having more resources, could afford them. As a result the poorer farmers were squeezed off the land that they had owned, and so were forced to join the growing populations already living in adjacent shanty-towns.

There has been an even more disastrous development. In the early 1970s the market prices paid by the West for cash crops that the Third World was able to export were relatively high, and repayments required by the West on its loans were relatively low. The difference between the two amounts was considerable, and it represented much-needed money flowing from the West to the Third World. By the end of the 1970s the prices paid for Third World commodities had become reduced, and the interest required on loans increased, to a point at which the two amounts were approximately equal. During the 1980s these trends continued until, now at the end of the century, the flow of money is from the Third to the First World, and to the remarkable extent of some $40 billions per annum. The trading exchange has come to be tilted on an enormous scale in precisely the opposite direction to what is so badly needed. The countries of Third World have thus been deprived of any obvious means of meeting their day-to-day necessities, or annual repayments of their debts, or of setting aside funds to finance their own self-development, and of hoping eventually to disengage themselves from the continuing dilemma that they face.

This situation is now, in its turn, having important secondary consequences. In Third World countries the struggle of the local people to survive has often forced then to extract from their land, forests, fisheries and so forth more than these environmental resources can satisfactorily sustain. The result is that their soils are becoming increasingly thin, overgrazed, subject to wind erosion and in some cases reduced to deserts, that their forests get cut down or seriously degraded, and their fisheries similarly depleted. Such underlying developments, especially if accentuated by drought or other setbacks, can easily lead to the kind of episodes of wholesale starvation that reach the headlines. All this bodes ill for the future. It is the climax of decisions taken by economists and men of commerce who probably lived in places far distant from those that are affected, and may have had little or no direct experience of the consequences of the accelerating downward spiral that has been set in motion by their decisions. The local people in the area know their environment and its problems from direct experience, and are often only too well aware of what is happening to their lives and living places. They are also aware that there is no possible way in which they, without external help, can hope to bring this disastrous downward spiral to an end. Careful early consultation between those who live and work locally and those who have the financial power, conducted in a way that would have led to an understanding of the essential problems, might have prevented much subsequent misery as well as unnecessary depletion of the limited natural resources.

(b) The environmental situation

The extensive human use of fossil fuel has marked a major turning point in the environmental history of the planet. When these fuels, whether in the form of coal, oil or gas, are burnt the potential energy that was associated with their creation many million of years ago (p.58) becomes available once more. An ability to use this energy greatly increased humankind's power. However its large-scale use proved also to have important drawbacks. The burning of coal, for instance in a power station, involves bringing once more the carbon dioxide and water that plants had originally used for photosynthesis (p.34) into existence. Consequently a power station ejects very large quantities of carbon dioxide into the air, and some of this finds its way into the Earth's upper atmosphere. Its presence there prevents some of the solar heat that is re-radiated from the Earth's surface at a longer wave length from passing through this layer of atmosphere and thus leaving the Earth. We may therefore be beginning to experience on the Earth a process that has been aptly called the 'Greenhouse Effect'. If this process continues, we will be dealing not merely with a local situation, for its consequences will be felt Earth-wide. It will affect climates (p.27) everywhere, and in ways as yet largely unknown. It may also result in considerable melting of the polar ice caps, and hence in a rise in the sea level everywhere. Such changes will have a profound effect on the environments of all the plants and animals, including the humans, living on the planet. A continuation, or still worse an increase, in the present large-scale output of carbon dioxide into the Earth's atmosphere is therefore highly undesirable.

The role of agriculture is important in this context. During the Middle Ages this activity gave rise to a fairly benign relationship between the plants, animals and humans involved. However since 1945 it has, in the West, all too often been transformed into what is essentially a special and unfeeling type of industry. The acute shortage of food during the 1939-45 war led to official encouragement of maximum food production. In Britain, for example, government grants and bank loans were given to farmers with this end in view. The results have been remarkable. Tractors displaced the earlier horse-drawn ploughs, and agricultural procedures in general became highly mechanised and motorised. In the flatter regions of the country hedgerows that had from time immemorial provided a last refuge for wild plants and animals were systematically removed in order to provide larger fields that gave greater freedom to manoeuvre the new machines. Similarly the intense competition led to fewer very large farms displacing the former more numerous smaller ones. A further consequence was that farm animals came to be treated with little or no feeling, as though mere lifeless commodities. This is demonstrated by the conditions in the many factory farms that have been springing up in numerous countries in the West. For these living creatures the environment in which they spend their lives has certainly not improved. In addition, farmers now apply large quantities of pesticides, herbicides and fungicides to their land, creating fields where no bird can safely nest, or wild flower grow, and that are devoid of almost all life apart from the particular crop that is being grown there.

These are just two examples of the way in which modern industrial techniques and mass production are damaging the environment. Other important issues arising from pollution are (a) the partial loss of ozone in the upper atmosphere with the consequent increase in the amount of ultraviolet radiation that reaches ground level, (b) the lakes that are being rendered lifeless by acidification, and (c) the damage to large areas of temperate forest possibly as a result of so-called 'acid rain'. Important research into these problems has been undertaken, and attention has been given to devising regulations and advising changes in techniques and procedures designed to reduce or eliminate the damage that is done. All this has been essential, and has enabled the position to be kept under some sort of control. Nevertheless the pressures are such that the land, fresh water, sea and air continue to suffer too much industrial pollution, and overall this is probably still increasing rather than diminishing.

The availability of energy derived from fossil fuels encouraged the development of technologies that enable people to go almost anywhere on the surface of the Earth and to do almost anything when they get there. These developments, operating in the self-seeking atmosphere generated by the prevailing socio-economic system, is resulting, directly or indirectly, in the wholesale destruction of forests, of Third World environments, of marine life, and so forth. Also the very rapid increase in the size of the world's human population (see below) impinges on, and increases the pressures on, the environment at all points.

The environment tends to be thought of as in some sense a peripheral matter, and of no great consequence. Yet it is an essential part of our only home, the Earth, and should concern us greatly. It affects the kind of people we grow up to be, the quality of our lives and our enjoyment of the living of them. It is likewise crucial to the well-being of the non-human life that shares this planet with us. Much further effort, increased awareness and understanding, and also research, are needed, and it is particularly important to find effective ways of eliminating the underlying causes that led to the present damage. Current information on environmental matters is readily available from such organizations as Friends of the Earth (FoE) and the World Wide Fund for Nature (WWF).

(c) The world population situation

The various circumstances that have arisen since humankind became an important feature on the world stage have themselves influenced the size of the human population. Changes at any given time depend on the difference between mortality and fertility rates. Long ago, when all people lived by hunter-gathering, the limited food resources necessarily restricted population size, and the total world population was probably only a few millions. Presently, with the adoption of agriculture, far more food could be produced. The population therefore rose. Later, with the development of civilization, still more food became available. On the other hand the more crowded conditions presumably increased the risk of death from infectious diseases. It has been estimated that by Christ's time the human population of the world had reached about 100 millions.

Thereafter for a long time populations rose only slowly. However with the coming of the Industrial Revolution more goods became available in the West. In addition this period brought a better understanding of the nature of infectious diseases, of the manner of their transmission, and resulted in reductions in mortality rates, particularly among the very young. For a considerable time fertility rates changed little and the size of populations in the West increased rapidly. Presently, however, as prosperity increased, it became apparent that there was no longer any need to have large families; indeed the quality of family life could be improved by keeping them small. This became more practical with the increasing availability of contraceptives. Fertility rates were thus reduced, and to an extent which roughly compensated for the reduction in mortality rates. Population size in the West therefore once more became fairly stable, but now at a much higher level than before.

In the Third World events have run a different course. Research in the West has enabled measures to be adopted which reduced the effects of some of the tropical diseases that afflict many Third World countries. Mortality rates have therefore been considerably reduced. On the other hand the kind of difficulties already noted (p.154-156) have prevented any corresponding increase in the standard of living, and so fertility rates have remained high. The result is that the size of the populations of many Third World countries has been growing, and often at a progressively increasing rate. In some of these countries this increase now amounts to 31 per annum and is giving rise to a doubling of their population within about twenty years.

Thus overall the human population of the world is at present increasing at an explosive rate. In 1850 it was about 1000 million, in 1950 2500 million, and by the end of the century it is expected to be nearly 6000 million. At present about 90 million persons are being added with each passing year. This is taking place on a planet whose size remains unchanged, and the resources that it can provide for human use are necessarily limited. This cancer-like growth of human kind on the body of the planet is one of the most pressing problems that we humans have currently to deal with.

(d) The science situation

Some recent developments in science are very important. At one time (see chapter 14) astronomy, chemistry, biology and geology were separate compartments of science, whereas now they have also become interrelated aspects of a single whole. It is even becoming possible to explore the nature of brains, thought processes and memory, and thus to begin to understand the nature of understanding. These developments, when applied to our planet, has had a further consequence. It has provided a means by which evidence existing at the present time but derived from the past (e.g. volcanic rocks, minerals, geological strata, fossils and so forth) could be used to interpret more effectively the approximate nature of the successive transformations that have taken place in the past, and how these have led to the situation existing on the Earth at the present time. It has thus made possible the kind of overall picture that the earlier chapters of this book have endeavoured to portray. With the help of allied disciplines such as archaeology and history we now also have some understanding of the emergence of humans, their early history, and of how, necessarily in the absence of any awareness of the nature of the overall scene, they has been making the kind of impact that is noted in the later chapters. This stage of relative ignorance is now passing. We are beginning to be able to see the world on Earth, and our human place in it, as a single interrelated evolving whole. This provides us with an opportunity to plan the future in an intelligent manner. It therefore brings a real glimmer of hope for the future, for it may provide the kind of overall map that will enable us to convert into more fruitful channels the present rather dismal prospects offered by the economic, environmental and increasing human population aspects of the present scene.

Another major change during this last hundred years has been the consciousness of the limitations of science. Old certainties have tended to be replaced by probabilities, and new deep-seated difficulties have been exposed to view. For instance, one can never the know the position of an electron, for the photon that can indicates its position also shifts it away from that position. Again, we can tell by observation what proportion of the nuclei of a radioactive substance will disintegrate in a given time, but it seems impossible in principle to determine which particular nuclei will in fact undergo this change. Similarly we can apparently never know what happened before the Big Bang, or why or how the Universe resulting from that event had the kind of nature that it appears to have had. Given this starting point, scientists have ascertained the approximate manner by which it subsequently came to be transformed from that initial state to its present one.

There are also fundamental limitations of a different kind. Our brains create a representation of a thing we see, but this representation is necessarily the product of an interaction between our nervous system and the thing in question. This thing itself—for instance, the mouse 'out there'—doubtless does exist quite apart from the human or the cat who may be looking at it, but there is no means by which one can know it in this pure unadulterated condition; to this extent it, and likewise all other things, necessarily continue to remain unknown to us. Also the scientist necessarily employs his nervous system in forming concepts, designing experiments and interpreting results. He (or she) is in fact himself a part of the Universe that he or she is endeavouring to study, and a part of the experiment that he or she is undertaking.

These various limitations are real and their implications are complex. It was a mistake to suppose that the Earth was our oyster or that the barometer was set fair. It is we humans who are now in charge and it is we who will have to mend our ways. Much humility is needed, and also much concern for the interrelationships of the various aspects of things on planet Earth.

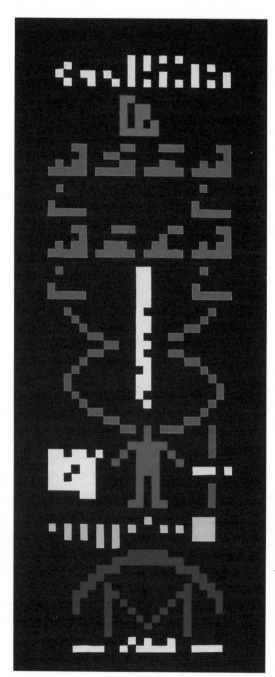

The Arecibo Interstellar Message. On November 16, 1974, a radio signal was transmitted from the Arecibo Observatory to the globular cluster M13, about 25,000 light years distant.

Arecibo Observatory; National Astronomy and Ionosphere Center, Cornell University

Chapter 16

THE NATURE OF THINGS: A SUMMARY

The previous chapters have provided a brief survey of the more relevant changes in the essential nature of things from the postulated beginning with the 'Big Bang' up to this present time. It seems important that we humans should have some awareness of this course of events, and of their implications, for it is only in the light of a background understanding that we can hope to deal effectively with the serious problems that currently confront us here on planet Earth. With this in mind I have tried in this chapter to indicate in a more precise form than previously the principal events that have led to the present nature of that planet and of the forms of life that inhabit it being as they are today.

This world in which we humans find ourselves as we grow up is a place with many aspects. We can see and feel them. There are, for instance, soil and rocks and hills, and also clouds and rain and streams. There is a thing called the Sun which passes overhead each day and warms and lights the scene. There are plants, apparently of many different kinds, the big ones being called trees. There are also animals, again of many different kinds, for instance 'blackie rabbit' and squirrels, and various kinds of birds—and they can fly—and also various less noteworthy creatures such as slugs and snails. There are also, and somehow of a rather different category, such things as fields and houses and electric light. Also, there are other human beings as well as oneself; however 'I' of course am 'I', whereas these others are merely others. More vaguely because one could not see them, and hence perhaps a little later, there are apparently things called countries; one was expected to support one's own country. And there was another thing that one did not see; it was called the Earth, or the world, and it was said to be round.

At first we take all this for granted. It is just the way things are. Yet it is complicated; there is so much diversity along with, apparently, so little rationality and, as we grow older, it all seems rather curious. The urge to understand the curious and so to make it seem less curious, to see how things are interrelated and so fit into an appropriate pattern, is apparently an inherent characteristic of human kind in general. And we—that is humankind collectively—have rather recently developed ways and means, known as scientific, which have in fact been helping greatly towards that end. The essential nature of the discoveries that have been made are of great interest in their own right, and are not unduly difficult to comprehend. A stage has now been reached when by the time we humans of the world reach maturity and have therefore become responsible citizens we should as far as possible all be aware of the basic understanding that has been acquired concerning the fundamental nature of this world in which we live, and of which we ourselves are of course one part. We would then be in a position to use our intelligence to make decisions and determine policies not merely as citizens of the world, but in addition as informed citizens who are aware of the underlying background of the problems that are being considered. This need, which is becoming crucial if democracy is to be effective, sets a major challenge to the educational systems prevailing in the world at present.

The evidence suggests that certain conclusions are well justified. Thus in the first place it seems clear that the Big Bang and its immediate consequences (e.g. space, time, energy, gravitation, electromagnetic fields) have to be taken for granted and accepted as 'given' since we do not know, and may never know, how they came into existence (p.9).

It is also clear that the condition of the Universe has not always been the same. Near the beginning, and thus a very long time ago, it was still quite small and exceedingly hot and there were as yet no atoms, or stars, and in short very little diversity. By a much later date, about 4600 million years ago, this Universe had expanded immensely, and at that time planet Earth came into existence as a small and apparently insignificant item within its total content. Here also the situation has been changing with the passing ages. There was a time when its land surface supported no life. Much later dinosaurs dominated the scene. And comparatively recently we humans came into existence, at first living as hunter-gatherers, and then, a mere ten thousand years ago, changing our lifestyle and creating for our own convenience from out of the natural materials of the Earth such new human-made systems as houses, fields and electric light. A youngster may, as noted above, correctly regard these as being in some sense different to the rest of his or her environment.

In the Universe in general, and not least on planet Earth which is the part of it that humans know best, the nature of things has therefore been changing in important ways. Furthermore this progression is towards the creation of integrated units, or systems, that become increasing complex and have in the process developed quite new kinds of qualities. This can occur because objects of all kinds, even one like a piece of iron that looks static and inert, is in fact composed of atoms each of which is a highly dynamic system (pp.19-22). If the form of these underlying systems changes then that of the object undergoes an equivalent change. For instance the atoms that compose the material called hydrogen are dynamic systems of one type, and those of oxygen are a different type. If these two systems are brought together their atoms mutually interact with one another in a way that creates a new third type of integrated system, known as a molecule. The result is therefore the creation of a new kind of substance, in this case water, whose nature and qualities are quite different to either of the parent substances that were responsible for its coming into existence (p.20). The result of such processes has been the creation of many new kinds of systems, at first non-living, but presently also numerous living ones. An evolution of diversity has been on the march on planet Earth.

New types of organization, and thus new kinds of things, can therefore come into existence. For any new system of this kind to emerge, to become plentiful, and to continue in that state for a considerable period, and thus prove relevant to the total scene, the following conditions appear to be required:

A. Origin Its origin requires the previous existence of appropriate types of system which can themselves, as a result of some form of reorganization, give rise to a new type of system.

B. Integration The constituent parts of this new system have to become well integrated, so that it becomes a single, effective and well-coordinated whole.

C. Environment The new system also needs to develop a satisfactory relationship with its environment.

D. Sustainability Such systems become more important if they remain in existence for at least some considerable period of time.

E. New overall situation The coming into being of a major new type of system is likely to cause important changes in its environment. And these in turn may interact with and further change this new system.

The above conditions probably apply to all the types of integrated organization. Using this framework, the characteristics of the more relevant new types of systems (T.S.) that have emerged are noted below, approximately in the order in which they originated. The intention is to indicate in a few words some of the crucial developments that have made possible the creation of the Earth along with its potentialities, and thereafter to consider similarly the subsequent changes that have taken place on this planet. It thus aims to provide a framework indicating the history, and hence also the present nature of things, on planet Earth.

T.S. no. 1   Hydrogen atoms

A. Origin It was the initial presence of protons and electrons, both of which were legacies from the Big Bang, that made possible the creation of hydrogen atoms at an early stage in the history of the Universe.

B. Integration A single proton and a single electron team up and respond to one another in unison, while remaining at a relatively great distance apart (p.12). This unified relationship is based on mutually attractive forces that exist between the positive charge carried by the proton and negative one by the electron. The two charges cancel one another, so that overall the atom is electrically neutral. The proton is far the more massive of the two units and so lies at the centre of the complex and is referred to as its nucleus. The electron, being far lighter, is the principal mobile partner in this system. The two units have thus come to form a single dynamic integrated system of a new type that is called an atom.

C. Environment For some time after the Big Bang the temperature of the newly-created Universe, and hence the degree of movement of the material units that were present, was so extreme that any hydrogen atoms that were formed were almost immediately disintegrated. In other words, their continued existence was not compatible with their environment. However after about 300,000 years the continuing expansion of the Universe, and the cooling that was associated with this, had reduced the temperature to a level at which the hydrogen atoms that were formed were able to resist disruption and they therefore continued to exist.

D. Sustainability Furthermore since atoms are closed systems their maintenance requires neither input nor output of energy (p.15); such systems can, in principle, just continue to operate forever. In normal environments these hydrogen atoms were therefore highly sustainable.

E. New overall situation In addition to protons, neutrons and electrons there now also existed in the Universe another type of system that was larger, more complex and of a different type to any that had existed previously. Hydrogen atoms thus represented a new and early creation in what is in fact an essentially creative Universe. It was a first step along the road to increasing diversity.

T.S. no. 2 Helium nuclei

A. Origin The raw material in this case was again that provided by the Big Bang, but in this case it was protons and neutrons, not protons and electrons, that were involved.

B. Integration This creation of a new level of integration also occurred very soon after the Big Bang. Normally protons repel one another because each has a positive electric charge, but under certain circumstances (p.12) they came to be exceedingly close to one another and then the so-called strong nuclear force, which was attractive and far more powerful, took over. The result was that two protons, along with two neutrons, formed a small, very compact and dynamic integrated system. Here again a new and more complex kind of system, but this time a nuclear one, had been created. It was a far larger and more complex nuclear system than the very different nucleus of hydrogen which consisted of a single proton.

C. Environment These nuclear systems, with two protons and therefore two positive electric charges, soon became associated at some distance from themselves with two electrons each with one negative electric charge. The resulting system, again electrically neutral, was a helium atom.

D. Sustainability These helium atoms were also very sustainable.

E. New overall situation The result therefore was the creation of atoms that were similar in principle to those of hydrogen, but more complex. Also the nuclei of helium atoms were far larger and more complex nuclear systems than the proton that formed the nucleus of a hydrogen atom. This new type of atom, as also its nucleus, was thus also well integrated, compatible with its environment and capable of continuing to exist for a very long period.

T.S. no. 3   Stars

A. Origin The next major development that needs to be noted is the creation of stars. The necessary pre-existing entities were the hydrogen atoms or, more strictly, hydrogen molecules (p.13) and the initial organizing agent was gravitational attraction.

B. Integration The result was the gradual aggregation of vast quantities of hydrogen gas that formed huge spherical clouds (p.13). As these clouds of hydrogen contracted they came to rotate at an increasing speed, while at the same time their interiors became exceedingly compact and hot. The heat caused the hydrogen atoms to disintegrate back to the primary condition of separate protons and electrons. Furthermore the heat was such that these protons gradually became converted once more into helium nuclei. This process resulted in the production of great quantities of energy (p.14). This nuclear energy raised the temperature, and so also the pressure, and this prevented the exterior of the star falling inwards on account of its weight. Excess energy was radiated from its surface into surrounding space in the form of heat and light. The cloud of gas had thus been transformed into the stable integrated type of system that we call a star.

D. Sustainability Stars are open systems; they cannot function unless they continue to receive sufficient energy to drive their systems. In this respect they differ from atomic nuclei or atoms which can just go on unfuelled forever (p.15). Stars therefore change during their lifetimes, and in ways that depend largely on their nuclear fuel situation. They therefore have life-histories. In their early stages most of this energy is produced by the conversion of the protons deep in their interiors into helium nuclei. When eventually this source of nuclear fuel becomes depleted other nuclear processes lead to the creation of more complex forms of atomic nuclei, such as those of carbon and oxygen. In addition the lives of very large stars terminate in a catastrophic phenomenon known as a supernova (p.16). First the depletion of nuclear energy provides a situation in which much of the outer portion of the star suddenly falls inwards towards its centre. The vast amount of energy created by this process creates samples of all possible atomic nuclei ranging upwards in weight from iron to uranium. This immense implosion is responsible for, and almost immediately followed by, an explosion that is so catastrophic that it expels outwards into surrounding space a large proportion of the nuclear matter that had been synthesised in the star during its lifetime. Here, in the cool conditions in interstellar space, these nuclei assemble at their peripheries appropriate numbers of electrons in the usual way, so forming corresponding new types of atoms. Stars therefore are mortal, though by our human standards their lives are remrkably long; for example, our own local star, the Sun, has now been in existence about 4600 million years, and is at present probably about half way through the first and most stable part of its life-history.

E. New overall situation How, then, did the creation of stars alter the overall condition of the Universe? It changed it from being a cold monotonous place filled with tenuous gas into one that contains the localised hot luminous regions that we call stars. Within the nuclear furnaces that comprise the interiors of these stars numerous new types of atomic nuclei, of increasing complexity, were synthesised. After disposal into the cool of outer space by supernova explosions these nuclei gathered round themselves a cloud of peripheral electrons the number of which was equal to number of protons in the nucleus; the resultant atoms were therefore electrically neutral. The overall result was that instead of the two types of atom, namely hydrogen and helium, that existed initially, ninety other types had also been created. Each of these atomic systems has its own particular characteristics; thus, for example, oxygen, carbon, iron and phosphorus are all quite different kinds of material. Stars, and therefore the occurrence of processes that occur within them, are numerous; in our own galaxy alone there are about a hundred thousand million stars. We know that our own local star is accompanied by planets circling round it, and this may also apply to many stars. Stars, and their consequences, have played a very important part in the overall developing scene.

T.S. no. 4   Planet Earth

A. Origin The creation of planets is a by-product of the creation of stars, and the planets of our Solar System were formed at the same time as our local star the Sun, namely about 4600 million years ago. The creation of a diversity of types of nuclei of atoms in earlier generations of stars played an essential part, and so did the equatorial disc of matter that would have been circulating round that star and which would have contained a sample of the resulting atoms (p.19). It was the presence of these latter which thus gave rise to the material diversity that came to exist on planet Earth.

B. Integration Later some of the various types of atoms that collectively gave rise to planet Earth mutually interacted with one another, forming numerous types of molecules. These chemical reactions further increased the material diversity that was present on the Earth. These resultant materials then began sorting themselves out (p.23), so forming the Earth's interior, surface crust, oceans and atmosphere, and thus the main large-scale features that now exist on this planet.

C. *Environment*   The radiation from the Sun warmed and lit the surface of the Earth. Also the rotation on its axis gave rise to night and day, and, owing to a tilt in this axis, its revolution round the Sun caused the changing seasons. Its relation with its environment therefore encouraged further creative development.

D. *Sustainability*   The Earth's prospect of retaining these amenities seems excellent. The Sun is likely to continue radiating much the same quantity of heat and light, and the Earth to continue revolving round this source of radiation in much the same way, for a very long time.

E. *New overall situation*   As regards the Universe in general, the creation of the Earth would have been of little consequence. But as regards the Earth itself, it was neither unduly large nor unduly small and neither too close to the Sun nor too far distant from it; also it had a substantial atmosphere, plenty of water, and a crust with considerable material diversity. It was, in short, a planet with potential.

T.S. no. 5   Molecules

Molecules come into existence when two or more atoms, which may be of the same type or more frequently of different types, mutually interact in ways that create from their assembled parts new and more complex dynamic systems that are similar, in principle, to those of atoms (pp.20-21). It follows that the pre-existence of atoms is required for the creation of molecules. Pairs of hydrogen atoms freely interacted to form hydrogen molecules at an early stage in the history of the Universe (p.13); it is these molecules that comprise the great clouds of gas that, as a result of continuing gravitational contraction, eventually become stars (p.17). Contact between atoms formed in interstellar space as a sequel to supernova explosions led to some molecules being formed there. However in the course of planetary formation (p.23) great numbers of atoms of various different kinds came into close proximity to one another in a cool environment, and it was then that a creation of molecules began to take place on a major scale. In addition, mutual interactions could take place between these resulting types of molecules, creating more new types of molecules and thus increasing still further the total range of material diversity. It is important that when atoms meet and mutually interact it is their outer portions, and therefore their electrons, that make contact and reorganise themselves, thereby creating molecules. The nuclei are shielded from contact by these electrons, and are not involved. The resulting chemical changes are milder and less energetic than the nuclear ones that take place for example in the interior of stars and in the nuclear reactors that humankind has recently been constructing on the Earth; these are far more energy intensive, and give rise to shortwave radiations that can endanger living cells. It is chemical changes, not nuclear ones, that are responsible for wind, rain and erosion, and for the various climates on the surface of the Earth (p.27), and also for the creation and maintenance of the living systems that have come into existence there. It is the heat produced by radioactive disintegrations of unstable nuclei deep in the interior of the Earth that has been responsible for tectonic processes, and thus for volcanic eruptions, continental drift and the building of mountain ranges.

T.S. no. 6   Living cells

A. Origin The Earth is probably the only planet in the Solar System where life currently exists. It is a form of creativity that is of special interest to us humans, not least because we are ourselves living systems. The life on our planet is based on living cells, and of these the prokaryote type, being at a bacteria level of organization, is relatively simple. Their origin would have required the pre-existence of molecules; as noted above, a diversity of types of these were created during the early formative period of the Earth's history; also the passage of ultraviolet radiation through the Earth's early atmosphere resulted in other highly relevant types of molecules being synthesised (pp.30-34). As regards the date of origin of life, we know that stromatolites (p.33) were already present not less than 3500 million years ago, so presumably simple forms of life already existed considerably earlier. The Earth itself had come into existence about 4600 million years ago, so presumably life originated fairly soon thereafter, and thus perhaps about 4000 million years ago.

B. Integration One can suppose that at some stage a number of different types of molecules that were enclosed within a membrane became in some way able to coordinate their activities, so creating within their localised domain the core of what became a living system. This was a type of organization that was very different from any that had previously existed.

C. Environment These systems would have been very different from atomic ones; the latter were closed systems and required no intake of energy; they could just go on functioning forever. Living cells, by contrast, were open systems that needed to be constantly receiving energy from their environment, and therefore to be frequently adjusting their activities in relation to changing circumstances.

D. Sustainability These individual living units were delicate and vulnerable to all kinds of contingencies, and so were unlikely to live for long. This whole new living enterprise would doubtless soon have fallen by the wayside if these cells had not also been able to grow, and when they had reached a certain size to divide into two separate cells each of which was a copy of its former self. It followed that as long as their birth rate remained equal to or greater than their mortality rate, living cells could continue to exist. The death of individuals and its compensation by reproduction has been a major complicating factor, with both advantages and drawbacks, for all the subsequent forms of life, whether plant or animal, that have subsequently evolved from some such a small beginning.

E. New overall situation At first these living organisms would have been small and scarce, and the general scene on planet Earth would have been changed little by their presence. Nevertheless life now existed, and its potentialities subsequently proved to be remarkable.

T.S. no. 7   Photosynthetic cells

A. Origin The presence of fossil stromatolites in rocks not less than 3500 million years old indicates that already by that time living cells were using molecules of the complex substance known as chlorophyll as a catalyst that enabled them to use energy from sunlight to transform molecules of carbon dioxide and of water into those of sugar. They could then use this sugar to create most of the other organic materials needed for their living processes (p.34). Thus sunlight, chlorophyll and of course the cells themselves combined to make possible this photosynthetic process. Here no major new level of organization was involved; the prokaryote cells remained prokaryote.

E. New overall situation The result was that the prokaryote cells equipped with chlorophyll were no longer dependent on primordial soup; they could now grow, multiply, spread and live in all the waters of the world, but only down to the depth to which sunlight penetrated. Cells not equipped with chlorophyll obtained their energy by decomposing the corpses of these photosynthetic ones; they were thus putting raw materials back into circulation which could then be used again by fresh generations of living cells. A simple type of ecosystem had thus become established in the oceans. Also oxygen gas, a byproduct of photosynthesis, was being slowly added to the content of the atmosphere.

TS. no. 8   Eukaryote type of cells

A. Origin   There was a very long period, lasting some 2000 million years, when there appears to have been little fundamental change. Photosynthetic prokaryote cells continued drifting near the surface in the oceans. However about l5oo million years ago living cells were coming into existence that, though still single-celled, were considerably larger. It is probable (p.42) that these larger unicellular organisms originated as a result of a pre-existing non-photosynthetic prokaryote cell having ingested a photosynthetic one, and that the latter, instead of being digested, survived and came to cooperate with it. The two could thus have come to form a single larger conjoint cell, and thus a eukaryote cell, with its own new potentialities.

B. Integration   Perhaps the most important feature of these eukaryote cells was that different region within the cell became organised into a number of types of organelles each of which was engaged in some specialised form of activity. The molecules in these organelles were structured in ways that led to each type of organelle being able to perform one type of specialised work very effectively. The molecules in the nucleus, for example (p.37) were concerned with growth and reproduction; this now involved not only straightforward cell division, but also occasions when two individuals conjugated (p.38) which resulted in the genes in the new organism being reshuffled. A diversity of types of individuals within a species was thus created. One result was that the amount of variation arising within a eukaryote species was likely to be sufficient to ensure that at least some of its members would survive all normal types of changes in their environment.

E. New overall situation   Occasional conjugations thus rendered eukaryote unicellular species remarkably well equipped for becoming and remaining well adapted to ongoing changes in the environment in which they lived. It would also have helped them to integrate useful new mutations into their living systems. The emergence and evolution of eukaryote cells was a very important development in the history of life on Earth.

T.S. no. 9   Multicellular organisms

A. Origin   It was the existence of these eukaryote unicellular organisms that made possible the transcendence to the next higher level of integration, namely multicellular organisms. The eukaryote cell structure itself was retained, and so was the sexual conjugation with its reshuffling of genes followed by non-sexual mitotic divisions resulting in cells that all had the same genetic content. The difference in procedure that led to multicellularity was that the cells formed by these cell divisions did not part company and live separate unicellular lives, as in unicellular eukaryotes. Instead they remained in close association, so forming a single group of mutually interacting cells and thus an initial type of multicellular organization. All these cells would have had the same genes as that resulting from the conjugation. The fact that their earlier free-swimming ancestors were able to sense aspects of their environment, and to respond to what they sensed, would have helped eukaryote cells in this new situation to sense and react constructively to one another.

B. Integration   Subsequent evolutionary developments led to cells at certain appropriate positions in an embryo becoming able to undertake particular specialised activities. This was doubtless more efficient than the corresponding type of specialisation in unicellular eukaryotes, where it was specialised organelles within cells, not whole specialised cells, that were responsible for corresponding specialised activities. New systems, both hormonal and nervous (p.48,49), were developed to coordinate the different kinds of activity that were therefore taking place in the different organs of each individual, thus creating from its many separate cells a single highly integrated individual multicellular living system.

C. Environment   In the sea the planktonic plants had no need for multicellularity. These simple abundant drifting organisms provided abundant organic food for potential animal life. It may have been in response to this incentive that multicellular forms of animal life were evolved in the sea and came to prey either on the unicellular plankton itself or on other forms of animal life that had themselves been nourished by this plankton. The Burgess Shale (p.50,51) probably represents an early stage in this process, and this kind of situation will have eventually led to the diverse and abundant life and the complex ecological relationships that we know to exist in the sea at the present time (pp.52-54). On land the position was different. Here there was no life until at a comparatively late date (some 400 million years ago) unicellular plants developed a form of multicellularity, also based on eukaryote cells, which enabled them to occupy this very different type of environment (p.55). This potential source of food then enabled animals to follow.

D. *Sustainability* The conjugation process inherited from unicellular eukaryotes led (a) to diversity among the offspring, (b) to those that happened to be particularly well adapted to life in the environmental conditions that they encountered being the ones most likely to reach maturity and have offspring and thus (c) to a species as a whole becoming increasingly well adapted to the way of life in the particular niche in the particular environment in which it was living, and (d) to species similarly continuing to be adjusted to the slow changes in their environment that would be occurring over many generations. Specialization and competition were making living organisms increasingly effective practitioners in the business of living, though at immense cost in terms of redundant lives cut off short, and often brutally, in each succeeding generation.

E. *New overall situation* Thus a diversity of forms of multicellular life came to flourish both in the waters and on the land surfaces of planet Earth. Its principal source of dynamism was competition, but it also incorporated numerous localised cooperative alliances such as that between insects and flowering plants (pp.60-62) which helped both partners to deal with their respective competitive situations. Environment was of course being modified by this multicellular life, as also was this life by its environment. In addition the occurrence of mutations introduced innovations, and these greatly extended the scope for the evolution of increasing diversity.

T.S. no. 10   Environmentally-conscious organisms

A. *Origin* Here we are not concerned with a new level of organization in the sense in which this term has been used, but merely with the incorporation into a group of multicellular animals, namely mammals, of a modification that had particularly important consequences. To this extent it is comparable with the incorporation of photosynthesis into organisms that remained at a prokaryote level of organization. The pre-existing requirements were the evolution of mammals from some mammal-like reptiles and capacity of biological evol-

utionary processes in general to bring into existence new kinds of development, in this case more particularly that of a well-developed cerebral cortex as
an integral part of brain organization.

B. Integration This represents a further degree of specialization and integration within the brain, and therefore also of integration within the animal as a
whole.

C. Environment The result (p.69) was that mammals became able to form
within their brains some kind of mental image of the environment around
them, and of the problems that it presented to them. In addition, the inclusion
of memory into this complex meant that their contacts with their environment
earlier in their own lives would be remembered. They could therefore draw on
memories of their past experience to guide their future actions.

D. Sustainability After the dinosaurs became extinct these mammals were
able to make full use of this new intelligence potential. It gave to them a certain dominance over non-mammalian animals. It also helped them to evolve a
wide diversity of types of mammals (p.71), each of which was able to use this
newly-created intelligence to help it to become adapted to the particular ecological niche that its species now occupied. The prospects for the longterm
survival of animals of this type was excellent.

E. New overall situation This post-dinosaur fulfillment of mammalian
promise brought to the nature of life on Earth a new dimension. It now included representatives that were to some extent both intelligent and environmentally conscious.

T.S. no. 11   Self-conscious humans living in hunter-gatherer communities

A. Origin The origins are unusually complex. It is the group of mammals
called primates that is relevant in this context. Their way of life (pp.74-75) led
to the cerebral cortex becoming particularly large and with numerous specialised regions whose activities are closely interlinked. Three major features
seem to be involved. The first was essentially a continuation of the kind of
processes (no. 10 above) that had led to mammals in general becoming environmentally conscious. The second feature, shared also by some other groups
of mammals, was that members of a primate species usually live together in
small more or less cooperative groups. Thirdly, these cooperative mammals,
being particularly intelligent, now needed more advanced methods of com-

munication and inter-relationship with their environment. The overall result was the emergence of speech and its consequences (p.78), of the making of tools (p.76), and of the communal type of life that presumably developed at their home bases (p.79). And these changes would in turn have helped these primates to become not only environmentally conscious, but also self-conscious and socially conscious. The individual would, for example, have become aware of the relevance of his freedom of choice and his ethical responsibilities (p.84). As a result of such creative processes human beings essentially similar to ourselves had come into existence by some 40,000 years ago.

B. *Integration* As regards the resulting individual men and women the degree of integration would have been about as good as in ourselves. Their brains and other organs of the body would have worked together forming a coherent whole, and thus a single integrated person. As regards the small groups or societies in which these men and women lived, the degree of integration would have been much less and the overall position therefore much more fluid (p.81). The situation was therefore complex. There were no necessary restrictions on what an individual could do within his society, as there are for instance in the case of individual cells within an organ of a multicellular organism. However such freedom needed to be restrained. Anarchic behaviour would lead to the destruction of group living, and such living had long since become necessary for survival. Judging from hunter-gatherer communities living at the present time it seems probable that potential anarchy was prevented by a form of indoctrination. In each generation the elders of a tribe would have impressed on the youngsters the traditions of the community, and hence what was done and how it was done, and they would have undertaken this in a way that induced the youngsters to accept this traditional attitude to life without giving the matter much thought. This form of teaching would have helped to maintain the social integration that would doubtless have been badly needed. This was a cultural development, not an outcome of genetic evolution. It could therefore take place rapidly and would not be inherited. The process would need to be repeated in each successive generation.

C. *Environment* The people would presumably have been quite well adapted to the environment in which they lived, not least on account of the clothes and tools that they constructed, the shelters that they used or made and the advantages arising from group living. Also, again judging from hunter-gatherers living at the present time, the groups would have been well acquainted with the habits and distribution of the plants and animals that concerned them. They would also have been conscious of the need to avoid drawing from the

CREATIVE LEAPS THAT SHAPED THE WORLD

environment more than it could afford to give. Their survival depended on its welfare.

D. Sustainability These fully human animals continued to live by hunter-gathering for some 30,000 years, and could have continued doing so indefinitely. In fact, however, about 10,000 years ago most of them abandoned hunter-gathering in favour of an agricultural way of life.

E. New overall situation Self-conscious human beings had thus been added to the scene on planet Earth.

T.S. no. 12   Early agricultural communities

A. Origin The change from hunter-gathering was probably gradual, and in the Palestine region involved the discovery about 10,000 years ago that by sowing the seeds of wild cereals they were able to harvest a larger crop than if they had gathered seed direct from the wild grasses. They also found that they could tame herds of gazelles and goats by being frequently in their vicinity and providing them with some grain (p.87).

B. Integration To consolidate this agricultural practice the communities would necessarily have had to remain in one place; also, as food was now more regularly available, their populations tended to rise. Such settlements therefore flourished and became widespread. Each family would have cleared and subsequently managed their own particular fields, so there was probably less social integration and overall community spirit than in hunter-gatherer communities with their communal hunting and gathering and sharing of the products.

C. Environment In the cultivated areas the environment became greatly changed. The wild ecosystem was destroyed, being replaced by a few species of plants and animals that had been especially selected and tended by humans for human convenience. The genetics of these plants and animals was also changing, due to selective pressures imposed by people, not nature, on their reproduction. As regards the humans, their attitude would doubtless also have been changing; they would now be principally concerned with the cultivation of their crops and the reproduction and rearing of their domesticated animals rather than, in the case of hunter-gatherers, with the natural history and distribution of the plants whose berries they gathered and of the animals that they hunted.

D. Sustainability In these simple agricultural communities most of the relevant materials were being recycled, and provided they retained their small size and their simplicity they would have been able to continue to exist for a long time.

E. New overall situation Hunter-gathering and agricultural communities are both types of human societies, and the change to the latter did not involve a creation of any major new level of integration. Nevertheless the abandonment of hunter-gathering marked a crucial turning point in the history of the planet. The change made cultural evolution, not genetic, the dominant force. Thereafter it has been the thoughts and activities of the humans that, by creating new types of systems on their own account, have been the principal driving force. This cultural type of evolution was almost infinitely more rapid than the earlier biological one, for this latter depended on an interacting combination of gene shuffling, mutations, and natural selection. The new people-made systems impinged on the old prehuman systems. At the beginning of the change, for instance, a rather crude field would have displaced the natural ecosystem that had previously been present at the location. At the end, as represented by the present, perhaps the most extreme result is the vast expanse of monoculture, controlled by pesticides and herbicides, that has displaced the prairies where buffalo once roamed in the Mid-West of the USA. Thus cultural evolution can change the environment far more rapidly and completely than can biological evolution. Such changes in environment will of course change the form of the nerve network in the cerebral cortex of the youngsters who grow up experiencing it, and will change their outlook. These youngsters in their turn may further change the scene, and this cultural type of evolution can thus proceed apace with the changing generations.

T.S. no. 13   Early urbanised communities

A. Origin A creation of cities began in Western Asia about 6000 years ago, and thus in the same general area as agriculture had begun some four thousand years before. By that time most sites suitable for agriculture were being already cultivated and perhaps for this reason people called Sumerians set about clearing and draining portions of the difficult swampy land in the plain through which the Euphrates and Tigris rivers flowed before they reached the Persian Gulf (p.93). Much work was involved, but the result was the creation of land that was extraordinarily fertile, and furthermore this fertility was periodically renewed by fresh silt deposited by these two rivers in times of flood. At such places huge crops could be grown by relatively few people, and consequently the local population increased, becoming far denser than in earlier human communities. Some of these people, not being needed for agriculture, became merchants; they travelled to other communities that were still practicing agriculture on relatively infertile soil, where they traded some of their surplus grain for timber and for copper, silver and other metallic ores. These were brought back home and used by others who were no longer needed for agriculture and who thus presently became skilled carpenters, builders, metal workers and so helped to provide the various types of goods and services needed by these small urbanised societies.

B. Integration All this activity had to be organised. Craftsmen, for instance, had to be paid for their products, and these had to be transferred to the people who required them. It was the temple authorities who undertook this mass of organizational work and who thus created the type of integration on which the dynamism of these small city-states in Sumer was based. Also, to keep records and give receipts, they had to invent the process known as writing (p.95). It was the local god of each of these small new city-states who was supposed to be responsible for this economic organization; the temple authorities were merely acting as his agents. This climate of thought doubtless encouraged stability but discouraged criticism or unconventional activities.

C. Environment The natural environment would of course have been further disrupted by the emergence of these centres of civilization. Also the people in these cities would have largely grown up in a humanmade environment, not a natural one. In addition, the inhabitants became divided into separate groups in each of which the members would have spent much of their time engaged in one particular occupation. The work, the interests and the day-to-day environment, and consequently the type of organization developed in the members' cerebral cortices, would have differed from group to group. Different social classes would therefore have emerged (p.94). Thus these city-states had presumably lost the community spirit that would have been present in hunter-gatherer societies, and to some extent also in simple agricultural ones, where all the people would have been engaged in more or less similar activities and have had similar overall communal interests. On the other hand it improved efficiency by enabling each of the various types of goods to be produced by workmen who had specialised knowledge of their craft.

D. Sustainability This structural organization seems to have arisen merely in response to developing contingencies; it was not well thought out or well attuned to human nature. The concept of the role of each local god presumably helped to keep it in existence.

E. New overall situation These small Sumerian city-states were in fact frequently at war with one another (p.97). In Egypt a comparable civilization, based in this case on high fertility due to the annual flooding of the river Nile, was emerging (p.97). Trade between these early centres of civilization and non-urbanised areas tended to transform the latter into urbanised ones (p.98). Conflicts between the various resulting states led to the creation of larger imperial units, and these in turn were frequently at war with one another. In all these processes the human beings would, again, have been mutually interacting with the environment, now largely urbanised, that they themselves had been creating. It is interesting to note that at this present time we still have with us, though in a different and much larger form, separate sovereign states and competition and wars between them, and also social classes and various kinds of religious concepts. The way in which these early civilizations developed has probably done much to shape the subsequent course of human history. Cultural evolution was proceeding apace.

T.S. no. 14   Small Greek communities

A. Origin Scattered among or adjacent to these growing centres of civilization were various small communities that were as yet not greatly influenced by them. The Phoenicians (p.100) and the Hebrews (p.100) have been briefly noted, and the Greeks (pp.103-114) were particularly important. These last lived in scattered localities hemmed in by sea and mountains, so each community necessarily remained small. They were sufficiently distant from the major centres of civilization, such as Egypt, for their own development to proceed unhindered by them; on the other hand they were also near enough to be able to learn some lessons from them.

B. Integration Developments in the Greek community at Athens were especially interesting. Policy regarding its affairs was decided at meetings held at the Assembly at which all its male resident citizens were entitled to attend, speak and vote. The organization and type of integration of their community were therefore determined not by gods, temple authorities or pharaohs, but by the conscious decision of a major sample of the members of their own community. The Athenians had invented and put into practice a form of democracy, and this was something new to humankind.

C. Environment The Greeks had created communities in which the minds of the members could roam freely and constructively. Some of them became interested in the nature of the natural environment around them, and put forward views to account for its behaviour. The members of these communities actively participated in the development of their societies, supporting for instance the artists who were in this atmosphere creating sculpture, drama and architecture that was widely appreciated and enjoyed. All this was also new. Life was viewed as being for living and for doing things that were worth doing for their own sake (p.105).

D. *Sustainability* These Greek city-states were small, and adjacent civilizations were large and powerful, and after two or three hundred years the Greek states were overthrown. For a time some aspects of former Greek life continued to have an important influence on Hellenistic society (pp.111-114), but presently this also ceased to be significant. If these Greek city-states had in fact continued to survive and prosper, then the world might have become a very different place to what it is today.

E. *New overall situation* The Greeks escaped from the restrictions on thought and creativity imposed by the mode of life adopted by the early civilizations, and they were able to develop a new kind of zest for life and to explore the potentialities inherent in athletics, history, ethics, art and science, and thus in short in humankind. The awareness of this Greek initiative was almost wholly lost until its rediscovery by Islam a thousand or so years later. Here it was cherished and expanded, and later transmitted to and assimilated by the West and has thereafter played some part in its subsequent development.

T.S. no. 15   The present world

A. *Origin* Some developments that have taken place since the collapse of the Greek city-states and have contributed to the nature of the present situation are noted below as follows:

1 The creation of the Roman Empire, and the subsequent collapse of its western sector (pp.115-117).
2 The emergence of Christianity (pp.117-119).
3 The establishment of feudalism in the West, and the encouragement of education by the monasteries (pp.118-121).
4 The rediscovery of Greek culture by the Moslem world and its subsequent assimilation by the West (pp.121-123).
5 The gradual displacement of feudalism by commercialism as the main driv-

ing force in the West. Here human labour was usually in short supply, and the use of other sources of energy, particularly of falling water for grinding corn, was encouraged. The development of new techniques was also encouraged, and this presently resulted in, for example, the availability of printing, the use of the compass and other improvements in navigation, and the production of firearms (p.123-128).

6 This led to an ascendancy of regimes in the West over those of other regions of the world, and hence to the exploratory expeditions organised by Prince Henry of Portugal down the west coast of Africa, to the passage of ships round that continent and by that route to India, and hence also to incursions into that country and the destruction of the Moslem trading complex in the Indian Ocean. It likewise led to the exploits of Columbus and Magellan and the circumnavigation of the Earth (pp.128-130).

7 Somewhat later Galileo was developing a procedure by which the validity of concepts concerning the natural world could be checked by appropriate experiments, so making possible the subsequent systematic ongoing process of scientific investigation (pp.138-141).

8 Later the industrial revolution resulted in: (a) the mechanization of production, (b) much poverty, overwork and ill-health among the workers in the new factories, and a new form of class division, (c) the use of fossil fuel as the principal source of energy, (d) a further enhancement of the relative supremacy of the West over other regions of the world, and (e) often the partition of such regions among the countries of the West into zones in which each would have a monopoly of sale of the superfluity of cheap goods that it was now capable of producing (p.131-133).

9 Subsequently there followed the development of railways, of steamships, and later also of aircraft. People and goods, including heavy goods, could be freely moved from one part of the world to another. This world, once so much a place of regions far apart and each largely self-sufficient, was becoming more a single place whose regions, now less self-sufficient, were linked instead by worldwide trade and economics (pp.133-134).

10 Science had achieved a considerable understanding of the underlying nature of things on planet Earth, and was thus able to work conjointly with technology. This led to the creation of organizations of a new kind, namely industries that were primarily science-based, and which became able to manufacture, for example, new kinds of dyes, plastics and pharmaceuticals, and to make widely available such things as electric light, refrigerators, telephones, radio and television (pp.147-150).

11 Another feature is that our planet Earth, which does not increase in size, is now becoming congested as a result of the developments on its surface. Its human population is already huge, and it is apparently inevitable that it will continue to increase for some years to come (p.159). There are numerous areas, whether on land, in water or in the air, where pollution caused by human activities has now become acute (p.157). Stress caused by inadequate social or economic organization, or by lack of understanding, is present almost everywhere, as seen for instance by unemployment and its consequences, and by the general situation in which the Third World finds itself (pp.153-156). Economic and other strains have tended to encourage people in many countries to separate into ethnic, religious or other groupings. The members of such groups then tend to cooperate with one another with a view to improving the status or economic position of their group. The resulting confrontations frequently give rise to civil or interstate wars. There are, or were until quite recently, wars of this type in Bosnia, the Sudan, Somalia, Rwanda, Mozambique, Angola, Afghanistan, the Caucasian region, Burma and Cambodia, and likewise paramilitary activities in Palestine, Kashmir, parts of South Africa, the north of Ireland and elsewhere. All this is causing an immense amount of death and suffering, and much destruction of both natural and human-made environments, and has forced some twenty million people to become refugees. Another feature is that these warring parties require and will therefore pay for weapons, and the manufacturers of such weapons, mostly located in the West or the region that was formerly the USSR, are well pleased to engage in these profitable transactions. The overall result is that great numbers of people are killed or wounded, and the world has become awash with guns. Also developments in communication now allow financiers in London, New York or Tokyo to make transactions involving perhaps millions of pounds in the twinkling of an eye, and they often do so for private profit and without giving thought to the social consequences. Similarly modern transport now enables people and their equipment to be easily and safely transported to even the most remote places on the planet, and there, in these still relatively natural areas, they are able to wreak havoc in order to 'make' money, or to gain power or for notoriety. The logic of this system also seems to be defective. Capitalist economists assert that this system can only be healthy if it continues to grow. Yet it seems self-evident that such growth cannot continue indefinitely since the Earth, which is the source of the productivity, does not itself grow. Its size remains unchanged. At the same time we in the UK are told that international competition for business has never been fiercer, and that we in this country must pull our socks up. Doubtless the same thing is being said in other competing countries such as Japan, Korea and the USA. If all of us in our respective countries are thus trying to maximise growth in a finite world, then it is not surprising that competition has never been fiercer; and it is likely to become fiercer still as time proceeds. Also if we here in the UK, and likewise those others like us in Japan, Korea or the USA, all go pulling up their respective socks, then it is not clear how this action will benefit any of us in whatever country. In a sense the economic system is the key to everything, and with regard to this important matter humankind seems to have thoroughly lost his way. So much for sector 15A, namely the origin of the world condition in which we find ourselves today.

B. *Integration* All things on the planet are becoming interconnected. It follows that it is the activities on the total planet Earth that now need to be integrated. At present, however, the principle feature is competition leading towards chaos, not cooperation leading towards integration.

C. *Environment* The environment, both natural and humanmade, is becoming increasingly disrupted.

D. *Sustainability* The prospects seem very poor unless fundamental changes are introduced quite soon.

E. *New overall situation* The human world on planet Earth has become full. All activities and events are interconnected. In this setting the prevailing economic system, with its competition and its money-making ethos, is quite inappropriate. It tears society apart, preventing worldwide cooperation. And it tears the environment into shreds. Both the old prehuman dynamism of the world and its new potential human-made oneness are being destroyed. Indeed the state of affairs on the planet seems likely to become quite chaotic, and also irreversible, within a generation or two. There is an ironic contrast between this present situation and the potentialities that are now available for the creation of a new Earth-Human system that is well-integrated, environmentally beneficial and sustainable. The immediate problem now facing humankind is therefore the transformation of the present actuality into a rational, pleasant and sustainable planetary system.

So far this chapter has drawn attention to a number of occasions when things (i.e. systems) with important new emergent qualities have come into existence. Initially it referred to systems that led towards the creation of planets in general, and then to those that are relevant to evolutionary developments on one particular planet, namely Earth, since this is the planet that principally concerns us humans.

However the emergence of qualities or potentialities that did not previously exist would be of little significance if it did not have repercussions on situations well beyond their immediate boundaries. Consider, for example, living cells. They came into existence at a very early stage in the Earth's history, and were quite a new phenomenon. Each cell had its outer surface bounded by a membrane, and it was within the volume thus enclosed that the living activities of the cell took place. Outside the membrane was the mass of water that formed the cell's principal environment. However the separation afforded by such membranes was not total. One can visualise that even in the earliest such cells organic compounds from watery primordial soup (p.30) diffused through the membrane and provided the nutriment that the cell needed to sustain its activities. Similarly unwanted products resulting from those activities would have been excreted through the membrane into the surrounding environment. The point that needs emphasis here is that living cells and their environment are always mutually influencing one another's condition, the state of each therefore being modified by this process. The consequences of the activity of a cell therefore spread beyond its border. It follows that not only the cell itself but also its environmental consequences need to be considered conjointly for they comprise one single interrelated process.

The next new development that had wide consequences was photosynthesis. Presumably the complex molecule known as chlorophyll came, probably by chance, to be present in some living cells. This substance acted as a catalyst that enabled energy derived from the sunlight that penetrated through a cell membrane and thus entered the interior of a cell to facilitate chemical reactions between carbon dioxide and water in the interior of the cell which resulted in the production of the important nutriment called sugar. It also resulted in the production of oxygen which, being unwanted, was excreted. The unicellular organisms that employed this process could therefore now flourish in all the sunlit surface waters of the seas anywhere in the world, for they were no longer dependent on primordial soup. Those that did not employ photosynthesis also became independent, for they could now obtain sufficient nutriment by feeding on the dead or decomposing remains of photosynthetic ones. This latter process also resulted in the recycling of materials, so allowing this type of living process to continue. The emergence of photosynthesis thus gave rise to a very different kind of overall situation on planet Earth.

It may have been the gradual accumulation of oxygen that was being excreted in this way, and in particular the energy that was thus being made available to living cells by means of oxidation processes, that eventually led to the next major change in the overall situation. This source of energy would, for example, have made it possible for non-photosynthetic single-celled organisms to consume and digest, and thus feed upon, photosynthetic ones. There is reason to believe that it was an occasional aberration of this process, namely when a photosynthetic cell that was engulfed as potential food was in fact not digested, but remained alive, and when subsequently these two cells developed a symbiotic relationship with one another, that a single functional cell of the eukaryote type came into existence. This conjoint cell was sufficiently large and complex for the molecules within it to become organised into specialised groups, known as organelles, each of which was particularly well able to perform one special type of activity within its cell. One of these organelles, called the nucleus of a cell, housed its genes, these being maintained in a particular sequence by the chromosomes. This sequence is replicated each time a cell divides; it is also to some extent reshuffled when two cells conjugate. Outside the nucleus, and therefore in the cell's cytoplasm, there are various other kinds of organelles, such as the mitochondria that provide localised sources of energy that are needed at various points within a cell. These eukaryote cells were therefore much larger and more complex than there prokaryote predecessors, and they accordingly had a whole new range of potentialities. The complexity of things was continuing to increase.

It was some of these potentialities that made possible the rather formidable type of reorganization that gave birth to multicellular forms of life. Here each individual consists not merely of one separate cell, as had previously been the case, but was instead a product of the joint activities of many separate cells. Previously the series of cell divisions that followed a conjugation had given rise to cells that separated and swam away from one another and thereafter lived separate unicellular lives. The multicellular condition was presumably initiated as a result of such cells not parting company but instead remaining close together in a cluster, and thus in a position where the separate cells can be mutually sensitive to one another's state. Such cells presently become surrounded by some type of covering in which there were sundry orifices such as a mouth that allowed nutritive material from the environment to be brought into the interior of the animal in the form of food. Such clusters could therefore grow, with very large numbers of cells becoming involved, and with specialised types of cells being produced at appropriate locations, thus giving rise to different types of organs in the positions where they are required.

These organs are roughly equivalent to the organelles, but were now composed of large numbers of whole specialised cells within multicellular individuals, instead of merely specialised groups of molecules within a single-celled individual.

As regards the individual cells within a multicellular organism, each is still covered by a membrane, but its immediate environment is no longer a great volume of water, but of fellow cells within the body of an organism of which it is itself a part, along with the cytoplasmic fluid that surrounds these cells and all the various activities that are taking place therein. Each individual of whatever multicellular species begins its life as a single cell, namely a fertilised ovum, and this cell has then to initiate its own development in ways that will eventually lead to it becoming a very complex and highly integrated organism of the type characteristic of the adults of its own particular species. Various books by Richard Dawkins and others might lead one to suppose that it is the action of the genes within the nuclei of cells that are solely responsible for organizing the course of this exceedingly complex ongoing transformation from ovum to adult. However the scientist Brian Goodwin has emphasised that it is not the genes of an organism alone that perform this function; it is accomplished by mutual interactions between genes and various structures in the cytoplasm of the organism in question. In other words, it is at each stage the total organism, not just its genes, that is responsible for the course of its own development.

The resulting multicellular adults also mutually interact in many complex ways both with the physical and the biological aspects of their environment. The fossils of the Burgess Shale (p.50) provide evidence of the situation in the sea at an early stage of this multicellular development. And, starting from some such beginning, the seas and fresh waters of the world have now come to be inhabited by a great diversity of types of fishes, molluscs, crustaceans and so forth that have evolved there and have done so in accordance with the ecological relationships between them that have also been developing during this same process.

Life, and therefore all the developments noted above, were for a long time restricted to the sea or fresh water. Comparable developments on land had to await the emergence of multicellular forms of plant life (which was virtually non-existent on the sea) had come into existence and established itself on damp land surfaces. This did not occur until relatively late, about 420 million years ago. This plant cover then provided a potential source of nutriment that enabled a few species of multicellular animals to exchange the type of life they had been living in water for one spent instead at least partially on land. In this new setting the decomposition of organisms after their death renewed the fertility of the land's surface, and hence encouraged the growth of further generations of plants, and consequently also of animals. Life on land was beginning to gain ground.

Each location will have had its own characteristic daily routine. The sun rose, rain fell, or didn't, the local plants used energy from sunlight to manufacture sugar, and the animals went about their normal daily living. One might suppose that this routine would just go on with little change indefinitely.

Yet, over a longer time scale, the overall scene does greatly change. For instance the descendants of the frail plants that first struggled to maintain themselves on land did in due course evolve roots that were longer and stronger and better able to absorb water from the soil, and also strong branched stems that extended some distance above ground level and bore specialised leaf-like structures that were well adapted for photosynthesis. Later again, by now in Carboniferous times, such plants had evolved into great forest trees that were competing with one another to each obtain a sufficient share of water and sunlight. Also elsewhere other types of plants were becoming adapted in ways that enabled them to inhabit the drier and more difficult parts of the land's surface by evolving effective ways of avoiding desiccation, of dispersing their spores or seeds widely and so forth. The more successful of these devices led to the evolution of conifers (p.60) and of flowering plants (p.61), and it is largely plants of these two types that comprise the forests, woodlands, meadows, heathlands and bogs of the world today.

As regards animals, it was principally certain members of two major groups, namely vertebrates and arthropods, that made the drastic environmental jump from a life spent in water to one on land. For vertebrates, the first step was for early fishes to become amphibious, evolving into amphibians; these were still heavily dependent on water, especially during their young stages. To be successful on drier land further adaptations were required, and these were accomplished, though in different ways, by both reptiles and mammals. As regards arthropods, the most important group came to be insects. The simpler types, such as dragon flies (p.58), were early on the scene and spent (and still spend) their young growing stages in water; it is only the last stage that is equipped with wings, and this enables the adult females to fly over considerable areas of land in search of suitable streams, ponds or lakes in which to lay their eggs. The more sophisticated types of insects, such as butterflies, bees, wasps, ants and flies, evolved at a rather later date; in these the larvae also have become adapted to growing in some specialised way on dry land, and the adults, also equipped with wings, search out and lay their eggs in places suitable to the needs of their forthcoming larval offspring. Arthropods have in addition contributed some other quite important groups, such as spiders, mites and scorpions, to the life on land.

There are also somewhat comparable ongoing changes of a purely physical kind. For instance, people going about their daily occupations in Europe would not be aware, unless told about it, that, owing to continental drift (p.26), the distance between themselves and people in America was increasing by two or three centimetres with each year that passed. Yet in the longer term that same drift has created sufficient space between the two portions of what was once a single continent to hold the water of what is now the Atlantic Ocean. And that, by any standard, indicates a very considerable change in the overall world scene. So does the building of a new mountain range, for instance the Himalayan one, which has been, and in this case still is, being raised as a result of the buckling of rocks exposed to pressure arising from two continental blocks colliding with one another. Such emergence of new mountain ranges also has numerous secondary consequences. It alters geographies and climates. It leads to erosion of the mountains. And it may provide a succession of different biological ecologies each being characteristic of its own particular altitude, so demonstrating neatly the close interplay of biological and physical systems.

It is the web of interwoven connections of this kind, involving ongoing evolutionary interconnections between plants, animals and physical conditions, that has given rise to the overall nature of things during the long period when multicellular forms of life have been playing a dominating role in the living sector of the total scene.

However in addition to this background norm there have also been essentially incidental or chance developments, or combinations of these, that have led to major deviations from the above overall established norm. It seems, for example, that either very early mammals or their antecedents came to possess a special extra portion of their brains, the cerebral cortex, that enabled them to make an intelligent appraisal of those aspects of their environment that concerned them. As long as dinosaurs continued to dominate the life on land these mammals remained small inconspicuous creatures that would have appeared to be of no special significance. However 64 million years ago, and thus after a long such period, unusual worldwide events (possibly exceptionally violent volcanic activity, possibly the impact on the Earth of a very large asteroid, or perhaps a combination of both) led to the sudden total extinction of dinosaurs. Mammals, on the other hand, did survive this crisis, which gave them an excellent opportunity, accentuated by their cerebral cortex, to explore and take possession of the numerous environmental niches now left vacant as a result of the extinction of their former occupants. These intelligent animals therefore now became a diverse and widespread feature of the new overall world scene.

In the long-term the most important of these habitats proved to be high up among the branches of the forest trees where mammals known as primates became the principal residents. Here there was plenty of food, especially fruit and insects; also there was no great danger from predators. The need for these primates to be able to jump effectively from branch to branch resulted in their evolving particularly good coordination between hands, eyes and brains. This in turn enabled them to use their hands to explore, for example, the nature and potential edibility of the fruits borne by the branches where they lived. Also, as these primates lived in more or less cooperative groups, the results of such investigations were likely soon to be quite widely known. Thus by about 20 million years ago dryopithecine apes, which were then common in the forests, had come to have a powerful combination of abilities that were not available to most non-primate animals.

Later, about 5 million years ago, some primates of this general type moved away from the branches of trees down onto the ground. Here in this new and very different environment they found that some of the abilities they had evolved during long period of arboreal life served them very well in the new ground-based hunter-gatherer mode of social life that they now adopted, and all the more so because the bipedal method of walking that they adopted on the ground meant that their arms and hands no longer had to participate in locomotion, and so were left wholly free for use in other ways such as, presently, making tools or stone axes. In this new setting they also succeeded in developing an ability to speak, and in a way that could convey precise information to fellow members of their group. This intensive exploration of their own potentialities seems to have stimulated the evolution of a much larger brain that did become a very effective general-purpose organ. The period between about two million and one hundred thousand years ago was thus presumably a remarkably constructive and exciting period for our ancestors, being characterised by the development of numerous new interrelated abilities and achievements. By the end of this period these small groups of now essentially human hunter-gatherers would, at least in their African homelands, have been making a strange and unpredictable impacts on the prevailing, long-established types of ecological orthodoxy.

More recently, about 10,000 years ago, perhaps the most crucial change that has at any time taken place on planet Earth began, namely the forsaking of hunter-gathering in favour of agriculture. One result was that the site of every field that was thereafter being created was also the site of the destruction of a portion of the earlier very different type of environment that had previously existed there. Until comparatively recently this perhaps did not matter much, but by the end of the twentieth century houses, factories, roads or fields are sited almost everywhere, and the older order of things is being almost totally destroyed. A single-species aberration has thus displaced the previous norm, and that species, namely ourselves, is only now beginning to understand the situation, and has as yet done nothing to replace the previous norm by some new, satisfactory and sustainable one. The result has been the kind of developments indicated by the previous chapters 9-14, and, as regards the more recent past, by chapter 15.

This present chapter has endeavoured to summarise the approximate way in which the changing nature of things on this planet seems to me to have evolved. The situation now in 1997 appears to indicate that unless the current drift is successfully altered as a result of our now having a better understanding of the situation, the state of this planet will quite soon be reduced to a highly obnoxious state of total chaos. Furthermore, this will have been brought about entirely by us humans. Unfortunately so far most of us still remain largely unaware and unconcerned, and so continue to fiddle comfortably while Rome is beginning to burn.

Chapter 17

OURSELVES AND OUR ONLY HOME

What then should we humans here on planet Earth do about trying to transform the present actuality into the rational, pleasant and sustainable planetary system that one can perhaps vaguely visualise? To determine this will he a major undertaking, and one in which perhaps all humankind should as far as possible participate. Here I can merely refer to some of the features that emerged from the previous chapters and which should, I think, receive attention in any such consideration.

1 Creativity

A very remarkable feature of the Universe is its capacity to create new kinds of things with new qualities, and hence to give rise to new overall situations, out of pre-existing ones. Some consequences of this creative tendency that relate to planet Earth have been noted in the earlier chapters and are briefly summarised in the previous chapter. These consequences include the various categories of things (i.e. of systems) that currently exist on this planet, and these of course now include us humans and our works. The humans particularly engaged in the study of such matters, namely scientists, have been able to demonstrate broadly how this has come about, and thus how this planet, as a result of its inherent capacity for transforming old kinds of organised systems into new ones, has passed through various stages and has thus come to be the kind of place is today.

One outstanding feature is that we humans, with our especially well-developed brains, have become able to understand something of the nature of these transformations and to create quite new kinds of things, but have unfortunately been doing so largely for narrow competitive purposes rather than for the good of a wider whole. This rather undisciplined procedure is now rapidly destroying the relationships that have evolved in the natural world of earlier pre-human times. In addition, this human intervention seems at present to be leading not towards a new and perhaps more appropriate situation but, on the contrary, towards overall planetary chaos. Furthermore it is only we humans, again on account of our especially well-developed brains and our capacity for understanding, who are potentially capable of rectifying this situation. There is now an urgent need for us to make a careful and sustained attempt to try to understand the nature of the problems that currently confront us and how we might use the creative potential existing on this planet, and not least within ourselves, to create a more satisfactory order of things here on the planet on which we live. It is in bringing to fruition this potentiality that our hope for the future lies.

2 Planet Earth

The Earth is one of a number of planets in our Solar System. These were formed as an integral part of the process that led to the creation of its central body, namely our local star the Sun. It seems in principle probable that numerous other stars have planets similarly associated with them. However as planets are cool bodies that do not themselves emit light it is difficult to recognise their existence unless they are relatively close at hand. Thus, since even the nearest other stars are immensely far distant from our Sun, we still do not know for certain whether comparable Solar Systems do in fact exist elsewhere.

A crucial difference between stars and planets is that the former are exceedingly hot and the latter relatively cool. Thus the two principal forms in which matter in the Universe exists in bulk are in a sense complementary. Stars demonstrate the results of creative processes taking place in high-temperature environments (chapter 2), and planets those occurring in relatively cool situations (chapter 3). Planet Earth is the most creative planet in our Solar System, and here such features as a solid portion, oceans composed of water, and an atmosphere were formed at an early stage. It thus became a planet with much further potential, and in this environment such complex and highly integrated systems as living organisms in general and, most complex and intricate of all, us humans, subsequently evolved. Possibly we are exceptional products of the Universe's creative tendencies. This also implies that we have exceptional responsibilities, for fulfilment of the great potential scope for further creativity on this remarkable planet now depends largely on ourselves.

3 Biological Evolution and Cultural Evolution

It is important to distinguish two very different types of evolutionary development. Biological evolution has been virtually the only process involving changes in life's forms from the time when life originated, not less than 3.5 thousand million years ago, until a mere ten thousand years ago. It has depended on such factors as cell conjugations, gene reshuffling, mutations and natural selection operating in the context of successive generations and functioning in overall environments that were themselves changing. It has usually been an extremely slow process. A period of one hundred thousand years has often led to no great change in the nature of the forms of life existing on the Earth. But the changes that were thus evolved were inherited.

The cultural evolutionary process that emerged on the scene some ten thousand years ago was quite different. It began when humans, each already equipped with a well-developed cerebral cortex as a result of earlier biological evolutionary processes, began to explore the possibilities of increasing their food supply by altering the environments, and thus also the life-styles, of amenable types of plants and animals around them, so using them for agricultural purposes (chapter 10). This new procedure altered the environments of the persons concerned, and so also their own life-styles and the patterns of activities developed in their cerebral cortices. On the other hand such changes did not affect their genes, and so were not inherited. Each generation of humans therefore had to adapt itself to its environmental circumstances afresh. The change in environment introduced by one generation meant that the next generation had to adapt to an environmental situation different to that to which its parents had become adapted. This new generation therefore differed from the previous one in its outlook, and so was likely in its turn to modify its environment further. So the alternating interactions between ongoing human generations and their environments has created a positive feedback situation. The cultural type of evolution has naturally been vastly more rapid than the biological one.

The first result was that agriculture soon virtually displaced hunter-gathering. Urbanisation was presently also introduced, and since then there have been many further changes principally pioneered by the West, so eventually leading to the present situation. The speed of this cultural evolutionary change, already rapid in any case, has tended to accelerate progressively, and there is probably now as much change in environment and in associated adult human personality in a mere one hundred years as there was earlier as a result of biological evolution in one hundred thousand years. Progressive acceleration can be dangerous, and the present situation is getting badly out of hand. Thus our main immediate concern is not with biological evolution, which is still continuing at its more leisurely tempo, but with the course now being taken by cultural evolution.

4 Human Nature

Before its birth a baby's development will be controlled almost entirely by its genetic constitution, and hence ultimately by the way in which its ancestors' experience of past environments has, acting through natural selection, gradually shaped that constitution; its own environmental experience will have been limited to that provided by its mother's womb. After its birth it is the external world around it that provides its environmental experience, and its cerebral cortex, already constructed under the influence of its genes, will organise the pattern of activity of its neurons in accordance with this new form of environmental experience. Thus both the biological and the cultural aspects of evolution play important parts in moulding the adult personalities of each one of us.

As regards the relevance of the biological component one can, for example, note that at an early stage in life's development (chapter 4) those prokaryote individuals that were well adapted for survival would have been the ones that were in fact most likely to survive and that genes that favoured this development would therefore have been selectively favoured and so become embedded in the genetic constitution of their species. One can further suppose that this selective pressure would have continued to operate throughout all subsequent ages, and that it is for this reason that a survival urge is now strong in living animals, whether human or non-human.

Similarly in unicellular eukaryotes (chapter 5) genes associated with conjugation would have been selectively incorporated into the genetic constitution, for this would have ensured the production of a diversity of types of individuals within the species concerned. This conjugation process has therefore been subsequently retained. Sexuality and the associated sexual urge presumably came into existence during the early stages of multicellularity (chapter 6), and thus somewhat later. The separation of the individuals of a species into two sexes enabled the sperm cells produced by a male to be small enough to swim through water, reach and conjugate with an egg-cell produced by a female, whereas these egg-cells, which did not need to be mobile, could therefore be large enough to contain sufficient nutriment to support the growth of the embryo that would develop from their union. Here also it would have been individuals in which the sexual urge was strong that would have been most likely to mate and produce offspring. This urge has therefore been retained, and is strong in virtually all multicellular organisms today. The present human position is complicated by the huge world populations, by the corresponding huge world gene pool, and by the fact that the total world human population has already become larger than the planet can satisfactorily support. The sexual urge remains no less strong, but fortunately its fulfilment can now be separated from reproduction by the use of contraceptives.

More recently the hunter-gathering phase of human evolution has also left its mark. It probably lasted from about the time of 'Lucy' and the makers of the footprints at Laetoli (chapter 9) some 3.5 million years ago until hunter-gathering was eclipsed by agriculture a mere ten thousand years ago. It was a period during which humankind's increasing tendency to develop greater brain power and capacity to speak and to make tools would have been mutually interacting, with advances in any one of these often leading to further development in the others. And in fact the relevant brains did increase enormously in size, and doubtless also in complexity and functional capacity, during this period. By about 40,000 years ago the pre-human condition characterised by 'Lucy' had become transformed into a fully human one; these now fully human individuals continued to live by hunter-gathering for some 30,000 years.

Speech, tool-making and much brainpower are still with us, and have all yielded abundant fruit. It is important that speech enabled individuals to attend to the needs and activities of their social group without themselves becoming physically embroiled in it; this left them free to develop a personal life and to form their own independent thoughts. Their position was therefore very different to, for instance, the cells in an organ of a multicellular organism, whose activities are totally dominated by the needs of the larger organization that they serve. This complex human situation still exists today and can lead, for example, to conflicting loyalties within the minds of individual persons.

One can also guess that during their long experience of hunter-gathering our ancestors came to inherit a certain empathy or accord with hill and dale, plant and animal, sunshine and stars, and thus for the natural world in which they still lived very natural lives. The consequences are still with us. Most of us enjoy walking through unspoilt countryside, seeing the plants and animals, feeling the wind and climbing the hills.

As already noted, the cultural component of planetary evolution, which is very different, has recently been getting into a confused and dangerous state. The overall result is that the world's most pressing problems are now of a social or cultural nature rather than a biological one. In a sense this is fortunate, for in view of huge size of the present human gene pool it would be virtually impossible to alter the genetic constitution of the overall human population and thus approach current problems from that angle. The cultural side of the evolutionary situation is however far more easily changed. The genes will duly organise a cerebral cortex, and its type will depend somewhat on the kind of genetic constitution that a particular individual has inherited. This is the position at the time of birth. Thereafter the very numerous cells of each such cortex will conjointly create patterns of activity that are in some way related to that individual's experience of his or her environment. The type of personality that the youngster develops as he grows up is at least in part a response to the mutual interaction that there has been between the activity of his cerebral cortex and the environment that he has experienced. And since he and his generation can themselves alter the environment, and this environment can itself alter the attitude of the generation that follows, the potential for rapid social and environmental change, perhaps even within a generation or two, is very great. The change in question can of course be either for better or worse, but we do at least in some degree have power to direct the overall nature of the change. The opportunities, and likewise the responsibilities, thus lie largely with us.

This emphasis on cultural evolution does not imply that its biological equivalent is no longer important. Problems related to plants and animals and their ecology, and to infectious diseases, genetic disorders, cancers and so forth are of a biological nature. They are very important, but nevertheless seem at present less urgent than those arising from the current social situation. Furthermore if these social problems were settled satisfactorily, then the biological ones could be researched, and results applied, in a more appropriate setting.

5 Science

We humans owe to science (chapter 14) most of our understanding of the nature of the world, including that of ourselves and of the planet on which we live. An effective procedure was initiated by Galileo and at first involved investigating one particular feature of one circumscribed aspect of the scene, in Galileo's case by measuring the rate at which bodies fall towards the ground. Generalisations about the matter are formulated on the basis of these measurements, and the validity of the ideas tested by further experiments. By allowing theory and practice to interact with one another in this way a stage will be reached when a view regarding the matter could be expressed with a considerable degree of assurance.

The employment of this scientific method of procedure presently also indicated that the concepts thus established concerning different aspects of the world were themselves interrelated and so were coming to form parts of a wider integrated network of human understanding concerning the nature of things. It also became apparent that pre-existing systems could on occasion interact with one another in ways that led to the creation of new kinds of systems, usually more complex ones, and so to kinds of things that were fundamentally new, and each with their own new natures (chapter 2). It was thus shown that the Universe was, at least to this extent, a creative place.

These scientific studies have also shown the approximate stages by which even before the Earth was born this creative Universe had been gradually evolving increasingly complex units from more simple ones.

Later, on the Earth itself, comparable changes have led to the creation first of unicellular organisms, then multicellular ones, then humans with their high level of intelligence, and now also the complex types of non-living organization created by these humans. The pre-Earth developments, mostly associated with physics or astronomy, are often so precise that they can be expressed in exact mathematical terms. On the Earth the complexity and the amount of variation associated with plants and animals, and even more with humans and their activities, has led to the precision and relative certainty associated with studies in physics being replaced in biology and more particularly in the humanities by less precise concepts that often therefore have to he expressed merely in terms of tendencies and degrees of probability.

The background understanding that is thus being made available by science plays a crucial role in technology, medicine and numerous other fields. It is also relevant to more incidental matters and more personal approaches. It has been found, for example, that there is a very high probability that smoking is a cause of numerous cases of lung cancer. However it is sometimes argued that no action should be taken until definite proof of this relationship has been established. The complexities involved in matters of this kind are such that absolute proof is often virtually unobtainable. In such cases there therefore seems good reason to take appropriate action at a stage of high or very high probability, rather than awaiting a degree of absolute proof that may perhaps never be forthcoming.

It is also sometimes argued that, for example, it is the size of the school classes, or that on the contrary it is the quality of the teaching, that is the cause of classroom problems. This somewhat confrontational either/or type of approach can sometimes be unhelpful. For instance in this case it may well be that both of the features in question are contributory, and that therefore both are in need of attention.

The relation between quantity and quality can also prove important. As was noted in chapter 3, if a piece of iron is heated it will for a time merely become hotter; this is a quantitative temperature change. However when the temperature of the iron reaches a certain point then, rather abruptly, the iron will melt, its former solid state being replaced by a liquid one. Here a change in quality has become apparent. Similarly, taking quite a different example, a few years ago it was widely assumed that the value of property would continue to increase more or less indefinitely. However, again rather abruptly, this ceased to be the case and the whole scene changed. In such instances there is presumably a certain critical point at which the continuing quantitative change results in a sudden reorganization of relationships between various relevant aspects of the total scene, and hence to an overall qualitative change. The whole scenario will then abruptly change. Thus things can turn out to be far less simple than they had perhaps at first appeared. It can, at times, be helpful to appreciate this.

One may also note the relation between, say, a person from Croatia and one from Serbia, or alternatively between a black person and a white one. In each case there will of course be differences between them, though those between individuals within each race are likely to be greater than those of racial origin. A second and probably more significant set of differences will be due to their having been bought up in different communities, and hence in cultural environments that give rise to different modes of thought, ideas about how things should be done, and so forth. It follows that neither of them should expect the personality of the other to be closely similar to that of himself, and neither should have reason to suppose himself to be 'superior' to the other. Any concept of ethnic cleansing would appear outrageous and obscene. Unfortunately however the nature of this background situation, though appreciated by science for some time, still remains unknown to most of the persons who are likely to be directly involved. Ethnic animosities (and rather similarly some religious and social ones) continue to exist. In short, we humans have failed to make widely known exceedingly important known facts. In such important ways the teaching of the world's children still remains very defective.

The respective roles of competition and cooperation are also relevant in various ways. Competition probably became important in living organisms firstly because, each individual being highly complex, there is usually some variation between the different members of a species and, secondly, because in each new generation usually far more individuals are born than can possibly survive. This situation provided the basis for Darwin's concept of the origin of species by natural selection.

Competition can also occur between members of quite different species, for instance between predators and prey. The former compete with the latter to obtain prey, and the latter with the former to avoid being preyed upon. Both parties tend to evolve highly specialised ways of life (and corresponding body adaptations) that are mutually interrelated; thus antelope have become equipped in ways that may enable them to run sufficiently fast to usually avoid being caught by predators, and leopards become able to run even faster, at least for sufficiently long to sometimes catch an antelope. Rather similarly, after the 'West' had produced atomic weapons the Soviet Union followed as soon as it was able. Thereafter throughout the Cold War period each side endeavoured to 'improve' these weapons by developments that countered or exceeded those being achieved by the other side. Such interrelated co-evolutionary competitive developments tend to promote one particular form of specialisation to a degree that is detrimental to all-round ability, and may prove unhelpful in the longer term to both of the parties concerned.

Processes that evolve conjointly are numerous but not always competitive. The mutual interactions between people and their environment at the time when agriculture was beginning to develop, noted in chapter 10, is one quite different example. Symbiosis is another, and here it is cooperation, not competition, that is the decisive factor. All such dual processes need to be viewed or studied in their totality since the essence of their nature is the frequent mutual interactions between their component sectors. It follows, of course, that both such sectors need to be considered or studied conjointly.

It is principally close cooperation between the activities of parts, not competition between them, that leads to effective and well integrated wholes. This clearly applies to the organelles that contribute so largely to the nature of eukaryote unicellular organisms, and similarly to the organs of multicellular ones.

Rather similarly, it is new types of cooperation between pre-existing systems or their parts that have been responsible for originating major new creative developments such as the first living cells, the first eukaryote cells and the first multicellular organisms, and indeed also for the emergence of photosynthesis, of the cerebral cortex, of agriculture and doubtless of much else. The emergence of a well-integrated planet Earth is needed, in which its human and non-human life, and the non-living aspects of the planet, cooperate, not compete, and so function in broad harmony, promoting the mutual welfare and sustainability of the various parts of this whole complex. This in turn requires the cooperative endeavours of the world's human beings to be focused on attaining this kind of goal, for it is necessarily we humans who will be the principal agents involved in the creation of any such fundamentally new level of organization. Unfortunately the present position is of quite an opposite kind. Past history has led to a situation in which competition, not cooperation, now reigns supreme. Different groups of humans are competing with one another to exploit the environmental resources existing on the planet; each doing so for their own particular and individual ends. This competitive approach is causing a rapid deterioration in the condition of humankind itself and likewise in that of his home the Earth. There is therefore now an urgent need to break away from this legacy from the fairly recent past. It has resulted in the course of the cultural evolution on the planet being determined principally by competition. We need to replace it by a situation in which cooperation reigns supreme. We humans have become fairly good at understanding existing situations and creating new ones of the type required to replace them more effectively, and we are now potentially capable of accomplishing the highly complex type of transformation that is needed.

6 Education

One could say that an early form of education was already coming into existence when the female ancestors of a fox, deer or other mammal was not only providing her offspring with milk and protection, but was also by her example indicating to them which kinds of situations were safe and which dangerous, and how the general business of their living could best be conducted. Such mothers were providing the cerebral cortex of their offspring with an early indication of the kind of situations these young would have to face later in their lives. This education enabled the latter to begin, at an early stage, to shape the development of their minds along the lines that would be required later in their lives.

Much later groups of adult hunter-gatherer humans, along with their offspring, would have used speech to plan their activities for the following day. At their living site the young would also be learning from their elders and beginning to shape their minds long before they themselves had to take full responsibility for their actions. Education had proceeded a stage further.

Later the descendants of these hunter-gatherers were actively transforming the nature of the environment around them, for example by creating agricultural systems, cities and sovereign states. The young had therefore now to learn and master a very different and far wider range of activities than previously. The range of interrelated matters that needs to be understood in principle by a responsible citizen has subsequently been continuing to expand, and it has now become very large indeed, not least on account of the recent advent of science and all that this implies. Clearly it is now impossible for educational curricula to pay attention to more than a small fraction of the aspects of this many-sided scene. In these circumstances it has become imperative that teaching should focus its attention on features that are relevant to a wide range of situations and will provide the pupils with some understanding of at least some of the more important aspects of the world into which they are growing up. It should thus also provide a foundation that will enable them to expand the range of such understanding more or less indefinitely during their later lives.

In this context it seems relevant that, at least in the more 'developed' areas of the world, there has been a tendency to select for inclusion in educational curricula certain 'subjects' such as geography, history, mathematics and science. What seems to be lacking, importantly, is an inclusive basic framework which would indicate how these various basic subjects are related both to one another and to humankind's present conception of the total planet. Something of this kind is now becoming possible, as I hope the previous chapters will have demonstrated. The present subjects could then easily find their natural place and interrelationships within this basic framework. This framework could include reference to aspects of astronomy, living creatures and other matters in which youngsters are likely to be spontaneously interested. Also these matters could presumably be illustrated by television and other teaching aids which would make their nature and significance more readily appreciated and assimilated.

In the longer term this kind of approach would hopefully give rise to populations in which numerous adults were eager and well able to live full and creative lives, and to do so within the context of a wider world to the development of which they themselves would be contributing constructively. Such communities would then hopefully cease to be merely democracies, becoming instead well-informed democratic populations in which a large proportion of the individual members would understand the basic interrelated issues that confronted themselves, their own communities and the world at large, and so would be able to draw conclusions and help to determine policies on the basis of the overall background situation. This would mark real progress, for a democracy whose people are without such understanding could be likened to a ship sailing without a chart and consequently with its sailors arguing blindly about what course to take. Shipwreck, not future progress, would most probably be its fate.

7 Future Prospects

What then will be the future of this spaceship planet Earth that is now in charge of these human animals that had earlier been created by itself? Among people in general there is at present a feeling that things are not going well either in their own particular country or on the Earth as a whole. The preparation of the last few chapters has seemed to me to indicate that this present worldwide malaise has been caused by developments along the following lines.

In the West (chapter 13), which has turned out to have been the crucial area, the initial seeds were sown when, between the 12th and 15th centuries, a commercial way of life gradually displaced the former feudal one. The economic relationships between people within a community then became based on competition between them rather than, as in feudal society, on cooperation aimed at attaining particular communal ends. Individuals then sought to become more wealthy. More effective ways of producing goods were searched for and encouraged, and improved techniques thereby introduced. New ideas or chance observations could sometimes provide the stimulus that led to quite new kinds of techniques being developed, and such activities presently resulted in, for example, such processes as effective ways of printing, the production of firearms and the use of compasses to guide ships when out of sight of land. These in turn opened vast areas of the world to exploratory and commercial activity by the West. Investigation of the underlying nature of things (i.e. science) was also encouraged, for it was appreciated that this might perhaps lead to the discovery of new useful ways of doing things. In general people thus tended to become more alert, able and effective and thus tended to give to the West a dominant and dominating position as compared with other regions of the world.

The Industrial Revolution followed as a sequel to these earlier developments. The cheap cotton that was then being grown in bulk in the USA as a result of the labour of Africans transported as slaves to America, the abundant energy made available by the use of fossil fuel and, lastly, the invention of machines that could automatically convert cotton into clothes, collectively made possible the mass production in factories of clothing that was vastly cheaper than any that had previously been produced. There was an immense potential market for these cheap goods, and not only in Britain itself but also in the less developed regions of the world that were at this time being colonised. In addition factories that specialised in the mass production of other types of goods soon came into existence. All these factory systems were based on much larger units, and were more capital-intensive, than any pre-factory modes of production. One consequence was the emergence of another new phenomenon, namely the creation of industrial towns that were built to serve the needs of these new factory systems.

These towns in turn gave rise to new scenes and problems. The combustion of the coal that provided the heat that generated the steam that turned the factory wheels also produced, as a side-effect, smoke and noxious chemicals that polluted both the towns themselves and the neighbouring countryside. Also the competitive and unregulated nature of the factories led to little attention being paid to the safety and health of the employees, to the overlong hours that they worked or to the scant wages that they received. These employees were therefore under constant strain. The quality of their lives was poor and their average life-span short. For a considerable time attempts to rectify such situations led to strong resistance, for the owners of the factories thought that this would reduce their profits and perhaps also upset the economic system that sustained this new mode of production.

Eventually, and continuing right up to the present time, there were created a succession of laws, regulations and codes of conduct that have been concerned with such matters as improvements in the machines that were used, and in the manner of their usage and maintenance. This has applied not merely to factories but also to the mechanised forms of, for example, transport and agriculture that have arisen as secondary consequences of the Industrial Revolution. Some of these continuing improvements have been remarkably successful. For example, travel by car is certainly more comfortable and probably safer than it was seventy years ago in spite of the vast increase in both the density and speed of motorised traffic since that time. Such changes have been necessary, and essential. However, importantly, such renovations have been almost wholly concerned merely with moderating symptoms arising from the underlying primary cause of these problems, and not with removing the cause itself, which was the competitive structure of contemporary human society. This is the driving force of this whole ongoing process, and it still remains fundamentally unchanged.

This prevailing economic system is therefore still intact and no less strong. And its long-term consequences are becoming more apparent. At present one sees the intensity of the competition continuing to increase, both between the various sovereign states and between individual persons within these states. Unemployment has greatly increased in numerous countries. In both the First and Third Worlds the rich have become richer and the poor poorer. Wars have become more numerous and so have the numbers of refugees created by these wars. The presence of small arms is becoming widespread within many populations. Damage to the natural environment is increasing. Also, importantly, there seems no good reason to suppose that increases of such kinds will not continue while the present socio-economic system continues to operate and ultimately to control all our destinies.

There is a general aspect of this general situation that may also soon become critical: the rapid increases of various kinds that are now taking place on the surface of a planet that itself does not increase in size. First, there is the present very rapid increase in the number of humans who inhabit the space available. This number is now increasing by about 93 million persons with each passing year, which amounts to about a quarter of a million in the course of each day. There has also been a rapid increase almost everywhere in the number of people who are unemployed. This greatly affects the lives of the persons directly concerned. It also has numerous side effects; in Europe, for example, it is encouraging a resurgence of ethnic antagonisms, which in turn is making it more difficult for the various countries to build a more united and cooperative European community. It is also making it more difficult for those seeking to escape from persecution to find asylum. The power of nuclear weapons, and perhaps now also that of chemical and biological ones, has increased to a point at which a major new war could lead to the destruction of large areas of the planet. Minor wars have become numerous and are devastating populations and environments on a more local scale. In addition countries not directly involved in wars are becoming increasingly reluctant to use parts of their limited resources to help fund the massive rehabilitation that is needed after such wars end. A rather different feature is the huge quantity of information that is now available and that can be transmitted by various new devices almost instantaneously to any part of the planet. This can easily give rise to misunderstanding, misinformation and other dangers in a fiercely competitive world. One might note also the greenhouse effect and ozone depletion arising from the continuing output of pollution, and nearer to home the numerous conservation problems, such as the increased ability of the various competitive fishing industries to decimate huge fish populations, of poachers to annihilate elephants, and so on.

If increases of this kind continue on their present course it seems reasonable to anticipate that fairly soon, for instance within about the next two generations, the present interacting network of activities will become so congested that there will be no scope for interplay or manoeuvrability between them. The whole world system will then lose its mobility, seize up and so grind to a halt. Some form of overall planetary chaos would then presumably follow.

This line of thought prompts three further comments. First, if we humans are to avoid this situation we will have to put appropriate counter-measures into effect fairly soon. Secondly, if we are to save ourselves we will have to begin, within this period, the process of transforming our present chaotic competitive social organization into one that is reasonably well integrated, fair to individuals and sustainable. Lastly, we are at present in no fit state to undertake any such radical reorganization. Much preparatory work of various different kinds will be needed first.

This same line of thought also suggests that four main types of scenario might emerge during the next fifty or so years.

Scenario no. 1 Early reorganization It is possible that a number of powerful decision-makers in multilateral and retail enterprises, banks, governments, universities and so forth might soon come to appreciate that if things are allowed to continue moving much further along their present course there will be catastrophic consequences both for the world at large and for the organizations whose courses they themselves are trying to steer. They may therefore try to adopt a policy that is strongly directed towards wider planetary issues as well as to the more localised ones associated with their own particular organizations. Developments along these lines could give greater stability to the world scene, creating a different perspective and a longer period during which more

208 ❖   CREATIVE LEAPS THAT SHAPED THE WORLD

substantial forms of reorganization could be planned, discussed and put into
effect.

Scenario no. 2   Later reorganization Developments continue much along
their present course until almost everyone comes to appreciate that a major
breakdown lies ahead. All minds would then be drawn towards this matter,
probably in a rather panic-stricken atmosphere. The resulting stimulus might
lead to some form of overall planetary reorganization that probably was not
well thought out. If this response were not forthcoming then one of the two
following scenarios could be expected.

Scenario no. 3   Some humans survive a major breakdown The   planet   becomes largely devastated, but some humans do survive and presently start to
reconstruct from the remnants.

Scenario no. 4   All human life is destroyed Some non-human life however
still survives. It might be expected to include tough and adaptable creatures
such as rodents and ants, and also some plankton and other relatively simple
forms of life. Biological evolution would start once more from this much lower
level of complexity.

The last two scenarios would mark, respectively, the partial and the total
failure of humankind to meet the challenge that now confronts us. To avoid
this situation by achieving either scenario no. 1 or some controllable form of
no. 2 will require early preparation. The purpose of the preparation will be to
create rather rapidly the widespread understanding that will enable the present
increasingly irrational and destructive world system to be replaced by one that
is reasonably well integrated, fair, creative and sustainable. To accomplish this
satisfactorily will require a widespread appreciation among people in general of
the essential aspects of both the living and the non-living components of the
Earth, and in particular of their basic relationships to one another.

The time has now come when we should view this small congested planet
Earth as a single ongoing whole, and thus also view its sovereign states, industries and indeed also its persons as integral parts of that whole. As with systems in general, the health of a whole depends on that of its parts, and that of
its parts on that of the whole; health also requires that the activities and needs
of the parts and the whole should be appropriately integrated. The need for
this type of integration would apply equally to both scenarios numbers 1 and

2. Also this preparation would need to start early because both the development of such understanding and its subsequent wide transmission will take time and, as already noted, time may well turn out to be in short supply. Thus 'early preparation' appear to be crucial words that may hopefully help to bring a satisfactory resolution to our current planet-wide dilemma.

People of different categories could make rather different kinds of contribution towards this preparation. Thus, as noted above, people of power and influence in industry, commerce or administration could pioneer the early redirection suggested by scenario number 1. Also, if they developed an appreciation of their own particular position within the overall planetary situation, they might contribute very effectively to the type of reorganization indicated by scenario no. 2.

Scientists could also contribute in various important ways. They are accustomed to dealing with systems, and this can help towards an adequate appreciation of the total scene and the consequences of various types of response. It might prove extremely helpful if they were to develop a quite new branch of science, 'Earth-Humanology'. This term refers to the co-evolving system in which the non-human and human components of the scene are continually interacting with one another and are viewed or studied conjointly. Its role would be to explore the whole field of Earth-Human relationships and to consider in this light how the ongoing welfare of both could best be achieved. Previous wide ranging assessments such as I have offered are necessarily only cursory but might provide a convenient starting point. They could be improved, extended, or totally replaced by more satisfactory ones.

The proposed new science could also consider, for example, the nature of the overall natural systems created on the Earth by biological evolutionary processes before we humans began to develop our cultural systems, and how these two very different sets of systems have subsequently been interacting with one another. It would also need to consider in some depth some of the problems here merely raised in passing, such as the extent to which the present socio-economic system is in fact responsible for many of the world's principal ills, what alternative systems might be introduced, and how a transition could be accomplished. Other important matters needing attention would he the possible future of the present sovereign states, and the nature of some overall planetary organization that could replace a United Nations which is at present essentially an appendage of these competing separate states. The eventual aim would be to consider the feasibility and demonstrate the relevance of creating an interrelated system of a quite new type that was based on an understanding of the total people-planet situation. This would be a more extensive and complex system than any that had previously existed, and would doubtless create various quite new qualities and likewise bring to view quite new future horizons. When satisfactory foundations for this planetary re-organization are firmly established it should be relatively easy to correct minor defects that experience will bring to light, so gradually making this newly reorganised world increasingly well-integrated and sustainable.

Teachers in schools and universities are also well-placed to make a major contribution. Much information is now available concerning the fundamental nature and inter-relationships of the various aspects of the world, yet most people have little awareness of its implications. A youngster may have had experience of such things as hills, streams, birds, humans, houses, food and electric light, but his or her daily routine will have provided little insight into how such aspects of the world mutually interact with one another. Yet this is the kind of understanding that will be essential if, now and later in life, the pupil is to play a full and constructive part in community affairs and in the wider scene on planet Earth. Much will depend on whether a youngster's teachers succeed in providing an adequate appreciation of the nature of such matters.

There are numerous non-governmental organizations (NGOs) which would almost certainly support developments of this kind. Most of them are mainly concerned with reducing the amount of damage caused by symptoms, and thus with such issues as preserving natural environments, assisting endangered species and helping living organisms, whether human or non-human, that are suffering from persecution, ill-treatment or neglect, or from the consequences of wars, or from the desire of particular humans to enrich or gratify themselves at the expense of other forms of life. Also, importantly, these NGOs assist by widely disseminating relevant information. They are widely and actively supported by the public, which incidentally seems to indicate that there is a strong desire within our communities for making a better world. A number of NGOs, in addition to dealing with symptoms, give serious attention to the underlying causes that give rise to these symptoms. In this connection I have in mind Oxfam, Friends of the Earth, Greenpeace, the New Economics Foundation and perhaps Scientists for Global Responsibility, but there are doubtless others. This conservation movement can both help and be helped by increasing recognition that welfare of the total planet should now provide the primary focus of all our attention and endeavours.

Last, but not least, is the vast general public to which we all belong. It mostly wishes well, but at present unfortunately has little means available to it of distinguishing crucial issues from incidental ones, or reliable information from non-reliable. This leads many people to suppose that they can play no useful part and consequently to become apathetic. Here also welfare of the total planet might provide a foundation that would help to clarify the issues and give rise to better-informed democracies.

There would also of course be interactions between these or similar enterprises. Thus the activities of the scientists working on the postulated Earth-Humanology project, of the teachers and students in schools and universities, of the various relevant NGOs and so forth would be contributing to a positive feedback situation that would lead towards the creation of an integrated Earth-Human type of overall organization. This process should be providing an opportunity for individuals to lead their lives to the full, and to do so within social and environmental situations whose potentialities they themselves had been trying to develop to the full. It would thus hopefully generate a zest for both the personal and social aspects of living that was somewhat comparable to that of the early Greeks (chapter 12); however this would relate not merely to one or two minute city-states such as the Athens of that time, but now to the whole planet.

A journey towards this postulated somewhat idealistic situation will certainly not be easy. It would have to start from where things stand at the time in question. The tabloid press would rubbish it. Many of the powerful institutions would oppose it. One would be told that embarking on such a course would have dangerous consequences; this might well be true, but the difficulty about alternatives that do nothing, or nothing effective, is that they would almost certainly result in much greater and more intractable dangers, for reasons that have already been indicated. To undertake some such journey does seem likely to be essential, and the project will have to have been subject to much thoughtful consideration and to have been widely understood and approved. Putting the eventual decisions into effect would doubtless be a lengthy ongoing process that would require much cooperation, patience and discussion, monitoring of progress, and perhaps above all personal integrity and perseverance. The journey would also be exciting and spiritually uplifting, not least because its accomplishment would be so clearly worth while.

It is this planetary transformation that will pose the principal problem. Once appropriate foundations supporting a well-integrated whole became firmly established, the future for the Earth and the various kinds of things on it would seem remarkably favourable. This planet is likely to go on turning on its axis and also travelling round the Sun in much the same manner more or less indefinitely. The course of the nuclear reactions that generate the Sun's energy will certainly bring the viability of life on the Earth to an end when the depletion of its stock of hydrogen nuclear fuel reaches a certain critical point. However this will not be for another three or four thousand million years; until then it seems likely that the Sun will continue to radiate much the same quantity of heat and light. There seems to be no life on the other planets associated with our own Solar System, and the Sun itself is such an immense distance from any other star (and therefore from any other source of planets) that it is exceedingly improbable that the situation on the Earth will be affected, whether for good or ill, by any form of life based elsewhere in the Universe.

The weakest link on the prospect horizon seems to be the nature of humanity itself, but once we get things running on an appropriate sustainable course it should be relatively easy for us to maintain its overall organization within the limitations needed to ensure continuing sustainability, while at the same time providing scope for the evolution of stimulating new creative developments. We can hopefully anticipate that if this present major crisis is successfully overcome both the living and the non-living aspects of the Earth will continue functioning appropriately—not quite for evermore, but nevertheless for perhaps the next three thousand million years.

In any case it does at least seem clear that this Universe, with its ability both to create and to destroy an immense variety of types of systems, is a truly remarkable phenomenon. We humans, who are one of the systems that it has created, have thus become one small part of its present totality. It also seems clear that our future in that Universe now depends on whether, before it is too late, we cooperate with one another and use our collective wisdom and energy to reshape the unsatisfactory relationships that currently exist between so many aspects of the planetary scene. This malfunction has been brought about inadvertently by us humans because we have tended to act independently of one another, each for our own particular ends and without regard to the wider consequences. It has become vital for the future welfare of all things on Earth, whether human or non-human, that we should now mend our ways and set about transforming this planet Earth, which is our only home, into the satisfactory, well-integrated and sustainable place that it is potentially capable of becoming.

Epilogue

SOME THOUGHTS ON FUNDAMENTALS

1 Human Nature

We humans are principally interested in humans, and more particularly in their welfare, both physiological and psychological. However we are not equally interested in the welfare of all humans. First on the list of each of us is usually 'Number One', oneself. Early in life we learn to say "mine, mine" and continue to give this precedence. This situation is deep-seated, perhaps universal, and may have originated at a rather early stage in biological evolution as a result of those individuals that felt a strong urge to survive being the ones that were in fact the ones most likely to survive and to bear offspring with this same urge. It follows that all through the ages and in all types of life those individuals that did survive usually had this tendency.

As one grows older one comes to appreciate that each of the various other persons around one is also principally concerned about his or her own particular 'Number One', and this of course is different to one's own. One also comes to realise that we humans cannot live in isolation. Our ancestors have evolved in ways that made continuing survival dependent on the different members of their group working together for its common good. This implied that the welfare of each of the 'number ones' in the group depended in no small measure on giving due consideration to the welfare of the others.

It also seems to follow that, the greater the distance, the more our feeling for our fellow humans tends to diminish. Closest in various senses are our families and intimate friends, then come others living in the same vicinity, then perhaps fellow nationals, and we are least concerned about those very many others living far away and perhaps under circumstances and with problems and outlooks about which we know next to nothing. Yet they, of course, have feelings, hopes and urges that are no less strong than those within ourselves.

How does the desire for sustainability fit into this complex human nature? As regards the individual it might appear non-relevant, for once he or she is deceased she or he can no longer know or care what happens. Yet the seeds of sustainability were probably implanted early, for the ancestors of those species that exist today must at all stages have paid adequate attention to the business of producing and rearing the next generation; this feature became embedded deeply in their constitutions. Much later, after the cerebral cortex had evolved in mammals, conscious awareness emerged in our human ancestors that members of generations as yet unborn will have urges, satisfactions and potentialities comparable with those that we ourselves experience. A conscious desire for sustainability may then supplement the pre-existing instinctive urge.

2 The Universe

The Universe has created within itself innumerable galaxies each with innumerable stars, many of which may, like our Sun, have planets circling round them. One of these planets, namely the Earth, has gradually created on its surface the life that exists there now.

What, then, is this Universe? How did it come to be? What purpose does it serve? Are such questions meaningful? I personally take a profoundly agnostic view as regards such matters. I certainly do not know the answers. I also think that no one else does. And I suspect that the answers to questions of this kind are in principle unknowable to human beings.

But this strange phenomenon that we call the Universe did come into being, and after it had had a second or so to settle from its initial highly hyperactive state it began to develop in a systematic way in accordance with 'laws' that seem to have been bequeathed it to and have to be accepted as 'given'.

It is this subsequent development of the Universe that we humans are becoming able, to some extent, to understand. Perhaps its most significant feature is that a succession of new kinds of things or systems have been coming into existence as a result of interactions between pre-existing ones. The new systems are usually more complex than their predecessors, and with their own new kinds of qualities. Thus within the Universe as a whole there has been a progression from the sole existence of such relatively simple units as protons and electrons to one that included atoms and some kinds of molecules, and also stars and planets. Then on planet Earth first living cells, then multicellular organisms, and then we humans were created in succession. Lastly, and rather recently, these humans have set about producing novel types of systems on our own account, including agriculture, cities, human organizations, technologies, sciences and so forth. One outcome of all this is our present perplexities and problems.

3 The Environment

The environment also contributes to the scene in important ways. The environment - the surroundings - involves by implication the entity that is surrounded. If this is some form of living organism it will need to grow, and to do this it will have to obtain nourishment from its environment. It will be drawing useful materials and sources of energy from its surroundings into its own living substance. It will also discard into this environment the waste materials created by its own metabolism. There will be a dynamic two-way interchange between organism and environment. Both will be affected by this process, but in interrelated ways; overall they form parts of a single co-evolving or coupled system which will need to be appreciated as a single evolving entity.

Relationships are usually complex. Aphids suck up the juices of a plant. So far the situation is simple. But a ladybird beetle comes to join the scene. For the aphids the actors on the environmental stage now include the ladybird and the plant; for the plant it includes both the aphids and the ladybird, and for the ladybird it includes the aphids and the plant. Their various mutual interactions will create a changing local scene.

Similarly, the attention of a cat is focused on a mouse. The mouse, deep among the grass stems, is from its different viewpoint concerned about the presence of the cat. The human so-called owner of the cat has strolled over to see what has attracted the cat's attention. The result will be that each of the three active participants will have their own particular viewpoint on the same overall environmental situation.

Each species therefore has its own type of interplay between itself and its environment. To some extent, and particularly in humans, this also applies to the separate individuals within a species. You and I have different genetic constitutions. We also have different parents, and the attitudes of parents have a very important influence during the early formative period of our lives. Also each of us has different environmental experiences, as a result of, among other things, living in different houses, doing different types of work, and having different friends and interests. One person views a forest as something to be cut down and transformed into financial gain. Another may live in that same forest, and love and cherish it. The plants and animals that collectively comprise this forest can play no part in determining their own fate.

4 The Global Scene

In the very early days, life on the Earth was restricted to various localised bodies of water in which primordial soup was plentiful. A major transformation may have occurred when just one or two of these unicellular organisms happened to create in the course of their metabolism an organic substance (chlorophyll) that was able to use the energy in sunlight to convert carbon dioxide and water into sugar and oxygen. The cells that consequently contained chlorophyll became able to use this photosynthetic process as a means of creating within themselves the kind of nutriment that all living cells require. These unicellular organisms were able to prosper and multiply until a stage was reached when they were living within all the sunlit surface waters of the seas everywhere on planet Earth. Life, and the associated life environment complex, was no longer merely local; it had become global. This resulting type of ecological network seems to have become integrated and stabilised, and thereafter to have continued to exist in this state, for a remarkably long time.

Eventually however there was a further fundamental change, again initially of a biological kind. It also may have been due to a single chance event occurring in the course of biological evolution, namely in this case a photosynthetic prokaryote cell becoming fused with a non-photosynthetic one, thereby creating a single conjoint cell that presently evolved into cells of the eukaryote type. This new kind of cell came to have a far wider range of potentialities than its prokaryote predecessors. These cells in turn prospered and multiplied. One of their potentialities was for cells formed by multiplication to no longer separate and live separate unicellular lives. Instead they remained in close association and their activities became so exceedingly closely integrated that they conjointly formed a single living individual of the new and far more complex type known as multicellular. These new multicellular organisms, each based on many eukaryote cells, came to develop their own new types of potentialities and living natures.

The photosynthetic activities of millions of millions of unicellular organisms continuing for perhaps 2000 million years had given rise to a high concentration of oxygen gas in the atmosphere, and also some dissolved oxygen in the water. This could interact with carbohydrates in some of the cells of these new multicellular organisms. This new type of respiration then provided a source of energy that these organisms could use to reach, grasp and consume other adjacent living organisms. They were thus becoming predators that were stealing and using the source of potential energy that their prey had, by their photosynthetic activity, been gradually accumulating for their own use. By some such process was created the predator-rich type of marine ecological system that the Burgess Shale and similar deposits show to have existed at a number of quite different locations in the sea some 530 million years ago. This new and more complex type of ecological system had therefore by then presumably also become global.

Since that time life in the sea has experienced a creation of many new species and an extinction of many old ones. Also overall there have been great increases in the numbers, the variety, and the complexity, and thus in the general richness, of marine life. Nevertheless since that far distant time there has been no really fundamental change in the global scene existing in the oceans of the world.

On the portion of the surface of the planet covered by land, not water, the course of events has been rather different. Living cells, always previously associated with water, could not readily become adapted to this more or less dry and very rigorous environment. Initial success seems to have come only comparatively late, between 450 and 400 million years ago, perhaps as a result of unicellular photosynthetic organisms becoming multicellular in ways that enabled them to meet the special requirements needed for life on land. The land plants that resulted, when once firmly established, provided an environment with sufficient food, shelter and shade to enable certain kinds of animals to leave water and join the plant communities on land.

The descendants of these early land animals have had a more notable history than their marine counterparts. It was presumably at some time in the very distant past that the unknown ancestors of the group of animals known as mammals evolved a new important external portion of their brains known as the cerebral cortex. Very much later, soon after the dinosaurs had become extinct, a subgroup of these mammals, namely the primates, became adapted to life among the branches of the trees in the abundant forests of that time. Here they made good use of the facilities provided by this cerebral cortex. It helped them to coordinate the actions of eye, brain, hand and foot as they jumped from bough to bough. It also encouraged them to use their hands to explore this environment, for instance by feeling the texture of the fruits and the nature of the insects that existed there, and discovering whether or not they were edible. This cerebral cortex also helped them to remember such experiences, to compare and inter-relate them, and to make use of what they had thus learnt.

Later some of these primates forsook the trees in exchange for life spent on the ground, where they lived as small cooperative groups of hunter-gatherers. Once more their cerebral cortex served them well, for it helped them to appreciate the nature of the very different environment that they now encountered. They became able, for example, to form a mental image of the various categories of things that existed in their vicinity, to express sounds, i.e. words, that represented these categories, and thus to create a form of speech by means of which the members of a group could communicate with one another with far greater precision than had previously been possible. They also found that they could use their intelligence to create quite new kinds of things, for instance by altering the shape of appropriate stones in ways that converted them into effective tools. Similarly they could create quite new situations using, for instance, the controlled use of fire.

Developments of this kind mutually interacted with one another and led to the creation of a positive feedback situation that enabled our pre-human ancestors to understand the environment in which they lived, and to make use of that understanding, to an extent that greatly exceeded the capacity of any other animal species.

Thus during the whole period between the origin of life and the beginning of a final crucial period just 10,000 years ago there seem to have been four really major transformations that changed the overall situation. These are the coming of: 1. photosynthesis, 2. eukaryote cells, 3. multicellularity and 4. the cerebral cortex. In all four cases the new situation seems to have originated in the course of biological evolution and rather possibly as a result of some single chance event. The duration of this initial phase would have been brief. The new system that emerged, if it were to prove effective, would have prospered locally and multiplied, so becoming more widespread. It would also itself be undergoing refinement as a result of the normal natural selection processes. At the same time this new type of life would both be affecting and be itself affected by the relevant aspects of its environment. The progress of this coupled evolutionary system would also be affecting the prospects and behaviour of many of the other forms of life that were already in existence. The planetary scene as a whole would therefore subsequently be involved in various complex ongoing changes. The tempo of these changes would gradually decrease as a new overall system became increasingly integrated and stabilised.

These various interacting co-evolving systems seem to have been at least partially instrumental in leading James Lovelock to refer to planet Earth as a living system that he has termed Gaia. The promotion of this concept has been valuable, for it has drawn widespread attention to the great importance of coupled evolutionary processes. However this planet does not grow in size and it produces no planet offspring; the requirements associated with the necessity for growth and reproduction, and hence also for food, lie at the heart of most of the fundamental characteristics of life on Earth. Thus, as I see it, planet Earth has given birth to the life here, and it has nourished that life and provided it with a home, yet, nevertheless, this planet Earth is not itself alive.

Finally there is that last and most revolutionary transformation which began about 10,000 years ago. It is still in progress, and is at present at a most critical stage. It did not originate from some new development in biological evolution, but from a further elaboration of the activities of pre-existing human cerebral cortices. Its first phase involved the replacement of the earlier hunter-gatherer way of life by an agricultural one. The most significant result was that in the cultivated areas the pre-existing system of ecological relationships, which had been gradually shaped and integrated by natural processes during countless earlier millennia, was now not merely modified, but instead totally destroyed. In its place we humans, aided by the experience of people/environment relationships gained by their cerebral cortices, had created a quite new kind of system and this, naturally, was now shaped and elaborated by themselves and for their own benefit alone. They were here creating a new kind of ongoing human/environment coupled system which naturally altered the post-natal nature of the humans as well as the environment.

About 5000 years ago this last major transformation entered a further phase. This at first involved the creation of small sovereign city-states, for instance in Sumer, which were continually competing with one another for goods, or for the means of acquiring goods, and were frequently at war with one another on this account. Today this basic situation still remains essentially intact, though in all respects now vastly larger and more complex. The states themselves are often vast, and so are their populations. These have vast quantities of energy, power and scientific and technological know-how at hand. The earlier wild or semi-wild environment is being destroyed, and nothing in particular is being created in its place. Aspects of many things are becoming increasingly interlocked, and the overall picture is chaotic.

The issues are no longer minute and local, but vast and global. Each of us needs to direct the power inherent in his or her cerebral cortex towards the task of conjointly attaining some appropriate form of global reconstruction. This endeavour is already beginning to take shape. Nevertheless the future welfare of life on planet Earth, whether human or non-human, seems at present exceedingly uncertain.

INDEX